ALSO BY JOEL ACHENBACH

The Grand Idea: George Washington's Potomac and the Race to the West

It Looks Like a President, Only Smaller: Trailing Campaign 2000

Captured by Aliens: The Search for Life and Truth in a Very Large Universe

Why Things Are: Answers to Every Essential Question in Life

Why Things Are, Volume II: The Big Picture

Why Things Are & Why Things Aren't: The Answers to Life's Greatest Mysteries

A HOLE AT THE BOTTOM OF THE SEA

The Race to Kill the BP Oil Gusher

JOEL ACHENBACH

Simon & Schuster
NEW YORK LONDON TORONTO SYDNEY

SIMON & SCHUSTER
1230 Avenue of the Americas
New York, NY 10020

Copyright © 2011 by Joel Achenbach

All rights reserved, including the right to reproduce this book or portions thereof in any form whatsoever. For information address Simon & Schuster Subsidiary Rights Department, 1230 Avenue of the Americas, New York, NY 10020

First Simon & Schuster hardcover edition April 2011

SIMON & SCHUSTER and colophon are registered trademarks of Simon & Schuster, Inc.

For information about special discounts for bulk purchases, please contact Simon & Schuster Special Sales at 1-866-506-1949 or business@simonandschuster.com.

The Simon & Schuster Speakers Bureau can bring authors to your live event. For more information or to book an event contact the Simon & Schuster Speakers Bureau at 1-866-248-3049 or visit our website at www.simonspeakers.com.

Designed by Paul Dippolito

Manufactured in the United States of America

1 3 5 7 9 10 8 6 4 2

Library of Congress Cataloging-in-Publication Data has been applied for.

ISBN 978-1-4516-2534-9
ISBN 978-1-4516-2538-7 (ebook)

For the Horizon 11—not forgotten;

and for all the people everywhere who do the hard work, unseen.

Contents

Prologue

It came out of nowhere, a feel-bad story for the ages, a kind of environmental 9/11. The BP Macondo well blowout killed eleven people, sank the massive drilling rig Deepwater Horizon, polluted hundreds of miles of beaches along the Gulf Coast, closed fishing in tens of thousands of square miles of federal water, roiled the region's economy, and so rattled the nation's political leadership that even the famously measured Barack Obama lost his cool, snapping at aides, "Plug the damn hole!" For months on end, the disaster seemed to have no quit in it. Admiral Thad Allen, the no-nonsense US Coast Guard commandant who did as much as anyone to keep the American people from losing their minds, said early in the crisis that the oil spill was "indeterminate" and "asymmetrical" and "anomalous." No one knew what that meant, exactly, but we got the gist of it, which was that this was a very scary situation that required very scary adjectives.

We were haunted by Macondo's black plume, gushing with lunatic fury on Internet news sites and camping out in the corner of the screen on every cable TV network. You could not escape the plume. It penetrated our psyche like a guilty feeling that won't go away.

When I told people that I was writing about the oil spill, they reflexively offered condolences, as though covering something so gross and repulsive and tragic must be an unending torment. But it was every bit as fascinating as it was horrible. Journalists are rarely given a chance to cover an event that is unlike anything they've covered before. Mostly we write the same thing again and again, with different proper nouns. There are formulas. There are templates. But this one had no predicate, and it caught everyone off guard. It burst from the murky water of the Gulf of Mexico late one night in the spring of 2010, too late to make the print run for the morning newspapers, and too unfamiliar in its details to trigger the immediate recognition that this would be the dominant event of the summer.

The disaster involved deepwater petroleum engineering, something most of us knew little or nothing about. We knew that oil companies drilled wells in deep water—somehow—but few of us had ever heard of a blowout preventer, or centralizers, or nitrogen-foamed cement, or bottoms-up circulation, or a cement bond log, or the danger of hydrocarbons in the annulus.

This story had its own interesting lexicon, a language crafted by men who use tools. Offshore oil drilling is rough stuff, hard-edged, coarse, and although there are women in the mix, they're few and far between. There is a heavy maleness even in the office jobs, in the cubicles of the company headquarters. A lot of the people in the industry are guys who got their education on the job, in the oil patch. What they do is complex, difficult, and dangerous. They drill holes in the pressurized Earth. They extract crude. They pump mud and cement, and handle gear weighing tens of thousands of pounds on a rig that weighs millions. Theirs is an environment dedicated to function, not form. And so even the language is masculine, the words often short, blunt, monosyllabic. Spud. Hot stab. Top kill. Junk shot. Dump box. Choke line. Kill line. Ram. Ram block. Ram packer. Side packer. Stack. Valve. Tick. Pod. Borehole. Bottom hole. Dry hole. Drill pipe. Coning. Cylinder gauge. Cavity. Rat hole. Reamer shoe. Wiper trip. Squeeze job. Squib shot. Stabber. Static head. Stopcocking. Torque tube . . .

A challenge in reporting the story was finding a way to translate Engineer into English. Those of us who covered the Marine Board of Investigation hearings—the joint inquiry into the Deepwater Horizon disaster by the Coast Guard and the federal Minerals Management Service (MMS), which regulates the offshore industry—heard witnesses say things like this:

> "At that time, we replaced all the rubber goods, the upper annular element, lower annular element, the ram packers, top seal, bonnet seals, then the riser connector ring gasket, 3/16th miniconnector ring gaskets, and the wellhead ring gasket." (Mark Hay, August 25, 2010)

And:

> "You've got some gas sensors on the mud pits. You also have a gas detector in the top of the shake where the mud goes into the shaker at the possum belly." (Jimmy Harrell, May 27, 2010)

In the following pages, I hope to turn the disaster and the struggle to plug the well into a tale that everyone can comprehend. The tragedy of the Deepwater Horizon offers lessons that can be applied to any complex enterprise: Take care of the little things. Pay attention to the stuff that doesn't quite make sense. Don't ignore those anomalies and hope they'll go away of their own volition. Respect the rules. Follow proper procedures. Don't ignore low-probability, high-consequence scenarios. Hope for the best, but plan for the worst.

The Macondo well blowout was a classic industrial accident, a sequence of tightly coupled events in which no single action could have caused the disaster. Some of the mistakes are screamingly obvious in retrospect, but at the critical moments, decisions were fogged by uncertainties. In this case, the critical hardware was a mile below the surface of the sea, where only remotely controlled vehicles could venture. People couldn't quite see what was going on. They literally groped in the dark. They guessed, wrongly—and people died, and the rig sank, and the oil gushed forth.

There was abundant human error in the mix here. Under oath, witnesses admitted that they skimmed key documents. They did not recognize that engineering anomalies were shouts of warning. They behaved as if past results were an accurate predictor of future events. They didn't take care of the little things, and then the big thing—the Macondo well, drilled by the Deepwater Horizon—didn't take care of itself.

The subsequent effort to kill the well was a white-knuckle enterprise. This was a technological problem unlike anything seen before. There was a hole at the bottom of the sea. And no one knew how to fix it.

The oil industry had been so successful in its expansion into the deep water that it had become complacent; it hadn't fully thought through how it would handle a deepwater blowout. The industry had failed to grasp that the migration to deep water would be a journey into a different world. In its initial, fumbling response to the disaster, BP tried to use the same hardware in the deep water that it had used on oil leaks in the shallows. The engineers didn't have the right tools, didn't have the right protocols; they were making it up on the fly.

The catastrophe echoed the Apollo 13 crisis of 1970. In both events, Houston-based engineers tried to improvise solutions to a novel problem in an extreme, inaccessible, hostile environment. But there was a key difference: The Apollo 13 crisis, in which a spacecraft explosion imperiled three astronauts on their way to the moon, lasted four days, while on Day 4 of the Deepwater Horizon disaster, the calamity was just getting rolling. Macondo was Apollo 13 on steroids, except when it was Apollo 13 on tranquilizers. Even as thousands of people scrambled around the clock in a frenzy of accelerated innovation, many of the deep-sea maneuvers had a glacial pace. The contradiction was unavoidable, because the well was full of unknowns, and the wrong move could backfire. Engineers feared that in trying to seal the well, they would incite an underground blowout that could let the entire oil reservoir bleed into the gulf. Decisions had to be vetted and fretted over by multiple teams of engineers and scientists. They were all sprinting in goo, running full tilt but hardly going anywhere, as if it were a bad dream.

The engineers found themselves trying to cram many years of technological innovation into a single summer. How do you contain or kill the oil gusher when the mere contact of methane and cold water at that pressure creates those damn methane hydrates that clog your pipes? And do it with the whole world watching, live, with Internet feeds of every mishap and blunder? While your company is suddenly a global pariah and your stock is cratering and every plugged-in investor is certain you're roadkill? With the media getting hysterical, and the president talking about kicking someone's ass? And scientists from reputable academic institutions warning that the oil spill will not only pollute the Gulf of Mexico but also ride something called the Loop Current all the way around Florida, up to the Outer Banks of North Carolina, and then onward to who knows where?

Apollo 13's rescue was a government operation; the oil spill response was an unusual and inherently awkward public-private partnership. The private sector had the tools and the legal responsibility for plugging the well, but the government had the ultimate authority for the response. That was confusing on its face. The public never understood the arrangement. Government scientists found themselves embedded in BP's headquarters, working cheek by jowl with BP engineers, and then the smartest man in the federal government parachuted into Houston along with a kitchen cabinet of freelance geniuses. Terawatts of brainpower were applied to the problem of the hole at the bottom of the sea.

What follows is a cautionary tale of a major engineering project gone hideously wrong, and the desperate effort to solve a problem that human civilization had never before faced. One recurring theme is that in an extreme crisis we should be thankful for the professionals, the cool heads, the grown-ups who do their jobs and ignore the howling political winds.

Another lesson to emerge is that in a complex technological disaster, hardware by itself won't solve the problem. You need to think things through, to diagnose and analyze and interpret. That can be a high art. A crucial breakthrough in this case happened far from the gulf, in an obscure government lab where an even more obscure scientist tried to understand enigmatic data points. In crunch time, call in the nerds as well as the cowboys.

You never know when someone's fantastically esoteric expertise may be called upon to help save the country.

Chapter 1

Macondo

On April 20, 2010, at half past one in the afternoon, Daun Winslow and three other oil company executives boarded a helicopter at the BP heliport in Houma, Louisiana, in the heart of Cajun country. They were heading offshore, to the drilling rig Deepwater Horizon. The executives wanted to meet with the top managers and engineers on the rig to discuss a number of safety issues, such as hand injuries, dropped objects, and the potential for slipping and falling. They also wanted to celebrate the rig's excellent safety record.

They would spend the night on the rig and return to Houma the next day, a twenty-four-hour round-trip. That, at least, was the plan.

Daun Winslow, whom investigators would later refer to in their notes repeatedly as "Don," worked for Transocean, the owner of the Deepwater Horizon. A lot of the men offshore have thick necks and barrel chests, and you can tell from a mile away they're good eaters, but Winslow didn't come from that mold—he was tall and thin, with a high forehead and deep-set eyes. He was marking this very month his thirtieth anniversary with the company. He'd gone offshore as a young man, armed with a high school diploma and a bit of college, and had worked his way up the ranks: assistant driller, toolpusher, offshore installation manager, and, finally, rig manager. The rig manager isn't on the rig, but is "in town," at Transocean's corporate office in Houston. Recently Winslow had been promoted again, to Operations Manager/Performance, North American Division. He was in charge of safety and efficiency on five drilling rigs in the Gulf of Mexico, including the Horizon. From the perspective of the Horizon rig workers, he was several rungs up the corporate ladder: the boss's boss's boss.

Some rig workers would later refer to Winslow and the other visiting executives as "the VIPs," or "the dignitaries." Offshore trips by the brass were routine, something the companies typically did on the third week of every month. The executives called this a "management visibility" trip.

Winslow was joined by his Transocean colleague Buddy Trahan, and by two BP executives, Pat O'Bryan and David Sims. BP had leased Transocean's Deepwater Horizon rig for nine years, and the partnership had been profitable for all concerned. The Horizon was a superstar rig with a sterling reputation. Just recently, it had set a world record for total depth of a well: 35,000 feet of casing in a Gulf of Mexico prospect named Tiber. The Horizon was a specialized machine for a specialized job, and it did not come cheaply. Transocean charged BP $525,000 a day for use of the rig. The actual tab for BP, once you threw in all the other costs—mud engineers, cement engineers, pilots of the remotely operated vehicles, caterers, the BP company men, plus the mud and cement, the food and water, and the basics of keeping a small offshore town operating around the clock—ran to more than $1 million a day.

Transocean had 139 rigs and ships, making it the largest provider of deepwater drilling vessels in the world. A Swiss corporation, it was the closest thing to a Swiss Navy. But Transocean, though a brawny corporation by almost any standard, was a sidewalk hot dog vendor compared to the mighty global enterprise formerly known as British Petroleum.

According to *Fortune* magazine, BP Plc. was the fourth largest company in the world, trailing only Wal-Mart, Royal Dutch Shell, and ExxonMobil. By certain measures, the company was richer than most of the nations on the planet. In just the first three months of 2010, it had reportedly made about $6 billion in profit. In 2009, even with oil prices far off their 2008 peak, the company reported $239 billion in gross revenue, more than the gross domestic product of Finland, Colombia, Portugal, Israel, Egypt, the Czech Republic, Singapore, Malaysia, Nigeria, Pakistan, or roughly 150 other nations. Though still rooted in Britain, and serving as a prime element of many a British pension plan, BP had evolved into the quintessential multinational corporation. The sun never set on the BP empire. It had become a dominant player in America in the 1990s, when it gobbled up rival oil companies Arco and Amoco. (For a couple of years, the corporate shingle read "BP Amoco.") It was now the largest producer of oil and gas in the United States, and the top provider of fuel to the Pentagon. The company sold roughly fifteen billion gallons of gasoline in the United States every year.

Big is what BP was, and big is what it wanted to be. Until the company's merger mania, it hadn't been much of a player in the Gulf of Mexico. Now it was the biggest fish in that small pond and had been aggressively acquiring leases to drill in the gulf. Steadily, hungrily, BP had ventured into deeper and deeper water.

As the BP helicopter, cruising at 8,000 feet, carried the executives toward the gulf, they could look out the windows and see, inscribed in the landscape, the history of oil drilling in South Louisiana.

The farmland and pastures below were growing scarce, the firm turf evaporating, the water gaining dominance. Here it came: marsh, swamp, bayou, bay. The landscape below could hardly be called pristine, because the hacking hand of human industry was everywhere. Before the invention of directional drilling, which allows the crew to drill to all points of the compass from a single rig, the oil companies had to park a derrick directly on top of their prospect. They cut channels in the marshes to haul the derricks to the drilling locations. Pipelines followed. The oil and gas industry went where it pleased. The result is a landscape of marshland that looks slashed, chopped, whacked—as if a madman or someone with a perverse antipathy to marsh grass went wild with a machete. There are also more blatant intrusions of modern industry: power lines, highways, shipyards, storage tanks, warehouses, refineries. Miles of green marsh and open water will terminate, abruptly, in an industrial zone.

In the 1920s and 1930s, as the industry really took off in Louisiana, the derricks sprouted in the middle of neighborhoods and backyards. The oil industry and the fishing industry drew from the same labor stock, and oil and fish became as inseparable as shrimp and grits. Young men raised as trawlers or crabbers or oystermen would chuck the waterman's life to start living in an oil camp in some godforsaken mosquito-infested swamp. There was excellent money to be made in oil even if you didn't have much schooling.

Even today, in a society that sniffs at a mere undergraduate degree, guys with only a high school diploma and maybe a little trade school under their belts can pull in six-figure salaries in the oil patch. These are good jobs, the kind of jobs the politicians promise to deliver. If you're offshore, you live half your life on the water—twenty-one days on the rig is a typical hitch for Transocean—and you're confined to a highly regimented, structured environment. But when you take that helicopter back to shore, you may well have a jumbo late-model pickup waiting for you in the parking lot at the heliport.

The helicopter passed Grand Isle, the last barrier island, shaped like a fishhook, with the brown-sand beach curling at its eastern tip into the rich waters of Barataria Bay. The 2010 calendar for Grand Isle was full of fishing rodeos. The redfish rodeo, the speckled trout rodeo—good times for charter boat captains and their clients who didn't care about fancy hotels but knew where to find game fish. Grand Isle, though not glamorous, is a sportsman's paradise. The presence of oil platforms right off the beach doesn't bother anyone. The oil platforms are like man-made reefs. Marine organisms grow on the pilings and pontoons, attracting small fish, which in turn attract bigger fish. The grassy islands and edges of the bay are prime nursery grounds for shrimp and crabs. Oysters abound.

Out the left window, the executives could see the final stretch of the Mississippi River, as natural as a catheter. The river's end has been re-engineered into an industrial object, a dredged shipping channel lined with levees, the banks firmed up by boulders. The river delta extends so far into the gulf it looks as if the Mississippi wants to flow to Cuba. This last stretch of river delta is known as the bird's foot, for the way it splays at the very tip, the river channel splitting into multiple outlets: Pass a Loutre, South Pass, Southwest Pass.

Then there is no more land, just open water. The northern Gulf of Mexico is dotted with oil platforms and lined with pipes that run along the seafloor. The continental shelf has been transformed into a machine—a vast, amazing apparatus—for the extraction and transportation of crude oil and natural gas. Nature has been conquered, bent to the will of human civilization. The engineers have shown that human ingenuity has no obvious limit. There's oil out there, and they're determined to go get it.

* * *

The geography, geology, physics, and chemistry of the gulf conspire to make it a terrific place to drill for the bewildering array of interesting molecular compounds that generically go under the label of "hydrocarbons." The geology could hardly be better suited for trapping oil in deep rock. There are abundant layers of sedimentary rock that have been deformed over millions of years into hump-shaped formations known as anticlines, which are natural traps for oil and gas. There are salt sheets and domes, remnants of ancient seas, that form impermeable caps on oil fields.

The oil and gas are natural consequences of the gusher of sediment and organic matter coming down the Mississippi River. The river as we see it today is of relatively recent vintage, geologically speaking, but the ancestral Mississippi—the original drainage of the continent into the gulf—began when dinosaurs still roamed the earth and there was not yet any such thing as the Rocky Mountains. The Gulf of Mexico is a rift in the earth's crust, 12,000 feet or so at its deepest point; maybe even a little deeper. The gulf began opening in the Triassic period, roughly 200 hundred million years ago, as North America separated from Africa and South America in the diaspora of continents. The gulf is continuing to grow; Cuba inexorably flees from Texas.

The gulf is an isolated sea, nearly walled off from the Atlantic Ocean by the combined landmasses of Cuba, Florida, and the Yucatan Peninsula. The gulf has always lacked the deepwater circulation of open ocean. Bad circulation means lots of anoxic layers, dead zones, places where there's so little oxygen that organic matter doesn't decay. That's great for the eventual creation of oil and gas fields.

The sheer volume of sediment deposited by the Mississippi River has one tricky consequence: Rapid layering of mud in deep, low-oxygen water leads to high rates of natural gas formation. Water gets trapped in the pores of rock buried quickly by successive layers of sediment. Imprisoned water increases the pressures in the rock formations. The Gulf of Mexico is famously "overpressured," as geologists put it.

Those pressures can lead to complications during drilling, and potentially a loss of well control. Thus, the gulf is a good place to find oil, and also a good place to create explosions. It is among the hardest places in the world to drill. The rock formations are gassy and unpredictable, and the major drilling prospects lie beyond the continental shelf, down the continental slope, within hailing distance of the abyssal plain.

The offshore industry dates back more than a century, to when the first derricks popped up in lakes and marshes and on piers jutting from the beach. Historians throw around different dates, but no one doubts that a breakthrough came in 1947 when the oil giant Kerr-McGee ventured past the Gulf of Mexico breakers and built the first platform out of the sight of land. There weren't many offshore platforms at first, and just a few hundred by 1960, but the government did what it could to encourage the industry, and eventually the western gulf was dotted with thousands of platforms that were in turn connected by tens of thousands of miles of pipelines. The pipeline network chased the platforms into deeper water, off the continental shelf. Some of the pipes carry the oil through the shallows and onto land, continuing onward to refineries with their twinkling amber lights, looking magical from a distance as they rise above the green marshes and bayous. Because the oil is piped directly on land, there is no need to off-load it from a platform onto a tanker and ship it to a port.

The upshot is that the view from a helicopter is different for an oilman than for a passenger who does not know what lies beneath the waves. The oilman senses the vast unseen apparatus. He knows that this network of wells and pipes is one of the wonders of the modern world—a petroleum paradise. But do people appreciate it? Do average Americans filling up their tanks know that the gas doesn't come from underneath the filling station? To an oilman, the American public and political leaders can be exasperating. They clamor for cheap oil but hate the industry that gives it to them. And so there are restrictions on where the industry can drill—an imaginary line straight down through the Gulf of Mexico, separating the rig-free east from the pipeline-networked west.

About an hour into their journey, the executives could see their destination through the helicopter windshield: the Deepwater Horizon.

Technically, the Horizon was a MODU, or Mobile Offshore Drilling Unit. Whatever you called it, and from whatever angle you viewed it, the rig was an awesome structure, towering over the waves in the middle of nowhere, in mile-deep water forty-eight statute miles from the last blade of grass at the tip of the bird's foot.

It looked vaguely like something you'd find on a NASA launchpad. The Horizon was 396 feet by 256 feet. People would ritually refer to the rig as being the size of two football fields, though the comparison misses the astonishing third dimension: all the massive hardware, the cranes, and, in the center of everything, emphatically jabbing the sky, the 242-foot derrick. This rig was, in effect, a giant hole-punching machine. Coming up through the middle of the drill floor was the top of the riser, the 21-inch pipe that, as the name suggests, rose from the wellhead far below. Racked one deck up were the hundreds of pieces of drill pipe that could be strung together to reach the bottom of the sea and then keep going for another two, three, four miles, deep into the rock formations, until the drill bit found the good stuff. The ancient treasure.

The rig carried more than four thousand distinct pieces of equipment: mechanical, electrical, hydraulic. It had six engines, 9,777 horsepower each. It had two 150-foot boom cranes, four 7,500 psi (pounds per square inch) mud pumps, four blowout preventers, two lower marine riser packages, six riser tensioners, two pipe packers, and seven linear motion and cascading shale shakers. Whatever these things might actually do, the point is, there was gear galore, everything you'd ever want if you were stuck in the middle of the deep blue sea and had been assigned the daunting job of creating, from scratch, an oil well.

The Deepwater Horizon had the remarkable ability to park itself on the high seas and remain almost perfectly stationary without resorting to an anchor. When drilling a well, it stayed so exquisitely positioned that the novice observer might suppose that it rested firmly on 5,000-foot pilings sticking into the seafloor. But it didn't. It floated. It was a *ship*. It could roam the gulf in search of hydrocarbons. Below the surface, sinking some 80 feet into the water, were two gigantic pontoons. Using eight 7,375-horsepower thrusters and all manner of computers and global positioning system readings, the thrusters took turns contravening currents and winds and keeping the rig right where it was supposed to be. Like any other ship, it had a bridge, where two dynamic-positioning officers took turns at the controls, monitoring the thrusters and the position of the rig. It rarely strayed more than a foot off location.

The rig had been built by Hyundai Heavy Industries in Ulsan, South Korea, and completed in 2001. Though owned by a Swiss company and operating in the federal waters of the United States, the rig sailed under the registry and flag of the Marshall Islands—a nation in Micronesia best known for having been blasted and irradiated by American nuclear weapons during atmospheric tests in the 1940s and 1950s. Having the Marshalls as the flag state saved Transocean a fortune in taxes.

The rig could accommodate 130 people. The arrival of the four executives brought the number aboard from 122 to 126—nearly a full house.

The Horizon was the jewel of the Transocean-owned, BP-leased fleet of deepwater rigs in the Gulf of Mexico. That was one of the things the executives from the two companies planned to emphasize during their meeting with the top rig workers. They would hand out the attaboys, and give the senior toolpusher an award for excellent performance. This rig set the standard for all the others. The Horizon could handle the hardest jobs in a very hard business.

But there was something the executives didn't know as the chopper settled on the helipad at two thirty in the afternoon, something they'd find out in just a matter of hours.

The Horizon had met its match.

Macondo

The weather was splendid: clear skies, no wind, no seas. The gulf was as flat as a pond.

The two top Transocean people aboard, Jimmy Harrell, the offshore installation manager (OIM), and Curt Kuchta, the captain, welcomed the executives. The executives signed in and listened to a one-hour safety lecture. Winslow had heard the spiel before and could have excused himself, but he listened attentively to make sure that the lecture included some of the company's recent safety provisions.

Each executive pocketed a card assigning him a lifeboat in the very unlikely event of an emergency evacuation.

The executives had arrived at a propitious moment. The Horizon had only a few tasks still to complete on the Macondo well. The rig had other jobs to get to, and was running about six weeks behind schedule.

Macondo was a hybrid well—exploratory, but designed to be a keeper, a production well. The Horizon had probed a formation, the M56, which BP geologists were quite certain held a commercially attractive quantity of oil. (There is oil all over the place, but the heavy capital costs of drilling dictate a search for the fattest, ripest fruit.) The Horizon had indeed drilled into exactly what it was looking for, and BP planned to announce that Macondo had been a major find, a reservoir of at least 50 million barrels of oil. Now the Horizon was pretty much done, having plugged the well with a thick glob of cement in a procedure of temporary abandonment. The Horizon would soon go on to its next job, and a different rig would show up to "produce" (the industry's way of saying "extract") the Macondo oil.

Reaching this point had been arduous for the company and the crew of the Horizon. Back in 2008, BP paid the federal government $34 million at auction for the right to drill in the nine-mile-square parcel of the gulf known as Mississippi Canyon Block 252. BP named the MC 252 prospect Macondo, after the star-crossed town in the classic Gabriel García Márquez novel *One Hundred Years of Solitude.* In a drilling permit request filed with the government, BP described the prospect:

> Macondo is a moderate depth Miocene prospect in the Mississippi Canyon area. The prospect is located entirely outside of any salt body. It is located approximately 24 miles north of BP's Isabela discovery which was drilled in MC 562 during 2006. The primary target for the Macondo prospect is the M56, which was the same as Isabela. The target depth for Macondo is approximately 18,400'. The well will be drilled to a [Total Depth] of 19,650' to test the older Miocene section below the targeted M56. Seismic data quality over this prospect is very good since there is no salt involved.

But the Macondo well had defied even the skilled workers of the Horizon. Things kept going wrong. They began referring to it as the "well from hell."

"Mother Nature just doesn't want to be drilled here," a rig worker,

Shane Roshto, told his wife, Natalie, when Shane was home in April and getting ready for his next hitch on the Horizon.

Macondo, like any well, was fundamentally a simple thing. It was a hole. The hole was wide at the top, near the gulf floor, and tapered to a mere 8 inches or so at the base, where a final string of steel production casing dangled in the hole. There were no pumps down in the well itself, no tricky machinery other than some valves and seals. The driving force of a deep well like Macondo is the earth itself: the weight of the crust of the earth provides pressure on the oil. The production of the oil takes advantage of the natural pressure differential between the reservoir and the surface. Engineers have techniques to help things along if necessary. They can, for example, separate the water from the oil after it emerges from the well and inject it back into the reservoir to maintain a high pressure at depth. But Macondo didn't look like that kind of well. It was going to pump itself. The ancient hydrocarbons, under high pressure in the deep rock, would flow upward in the direction of lower pressure.

In a sense, the oil and gas are predictable. The hydrocarbons never act out of character. They never defy the laws of physics. And those laws say that, without barriers, the gas will rise from the depths, expand, rise faster, expand more, and keep rising at greater and greater speed until *boom!* A blowout.

Which is the scenario that Shane Roshto, the rig worker, sketched out to his wife before he flew back to the rig: "It's like blowing up a red balloon and taking a pin and just pushing it and pushing it and pushing it as far as it could go and it just blowing."

A rig named the Marianas first started drilling in the Macondo prospect in October 2009, but a freak late-season hurricane damaged the rig and forced it into port for repairs. BP sent the Deepwater Horizon to take over the job. The Horizon began drilling in February 2010. The job proved to be problematic from the get-go. Every well has its eccentricities and challenges, and the Macondo prospect turned out to be not only gassy but also dominated by brittle rocks that would crumble when drilled. Such formations are known among petroleum geologists as "friable."

On March 8, 2010, something went seriously awry. The crew was drilling through a crumbly, friable area where the hole would continually collapse before the workers could send down steel casing to stabilize the well bore. The rig experienced a gas kick, a well-control problem. Next, the drill pipe got stuck. The crew couldn't budge it. Eventually the well team, with regulatory approval, made a strategic retreat, severing the drill pipe and leaving three sections of it in the hole. The Horizon then started drilling a new hole near the first one. The operation was, in industry parlance, sidetracked.

Worse yet, Macondo had shown itself to have an unseemly appetite for drilling mud. Mud is an essential tool in drilling, and the decisions on how to handle the Macondo well were shaped to a great degree by the mud losses suffered by BP and its contractors.

The mud is not really mud: It is a carefully engineered, manufactured product. Technically it is a "drilling fluid." It can be oil based or water based, and salt-water based or fresh-water based. It can be an emulsion mud or a surfactant mud. A common ingredient is barite, an element that allows for mixtures of greater or lesser density, given the demands of a particular well. Mud has been an essential element of oil drilling since the early twentieth century. Fully eighty-eight pages of the *Standard Handbook of Petroleum & Natural Gas Engineering,* volume 1 (William C. Lyons, editor) are devoted to mud, mud pumps, and "completion fluids." The mud has seven distinct functions, the handbook reports:

1. To remove rock bit cuttings from the bottom of the hole and carry them to the surface.
2. To overcome formation fluid pressure.
3. To support and protect the walls of the hole.
4. To avoid damage to the producing formation.
5. To cool and lubricate the drill string and the bit.
6. To prevent drill pipe corrosion fatigue.
7. To allow the acquisition of information about the formation being drilled.

So it is marvelous, multitasking, precocious stuff, busy as a beaver as it cools the drill, removes the cuttings, firms the hole, fights corrosion, and diagnoses potential problems for the rig workers. When the mud comes back to the rig, having circulated through the well, it speaks to the drillers—tells them what they're dealing with down there in the deep earth. If there are signs of gas in the mud, for example, the drillers will suspect that hydrocarbons have invaded the well bore and the crew should be on alert for a potential gas "kick," which at its worst can lead to a blowout.

Of mud's many virtues, none is more important than its ability to suppress gas kicks and blowouts. Mud is a barrier to hydrocarbon flow. The weight of the mud, filling long strings of pipe, counterbalances the upward pressure of the hydrocarbons. All that fourteen-pound, fifteen-pound, sixteen-pound-a-gallon mud—whatever that day's recipe calls for—is a boot heel in the face of a gassy well.

But the mud engineers, the workers who mix the mud from the various elements and chemicals, must have the fussiness of Goldilocks. If the mud is too heavy, it can cause downhole problems, such as fracturing the rock formation. Heavy mud can surge into pores in the deep rock and never be seen again. But if the engineer crafts a mud that's too light, too perky, it won't be able to stand up to the intense pressure from the hot hydrocarbons at depth. It'll get pushed around. The hydrocarbons can shove the mud right back out the well, and then the crew has a well-control problem on its hands.

Ideally the mud will make an uneventful round trip in the well. It'll go down, come back up, no issues. Macondo, however, was a well that liked to cause trouble. The heavy mud continually leaked into the porous rock of the ancient sandstones. It didn't circulate back to the rig. "Lost returns," the drillers call this (or, sometimes in more casual, Cajun speech, "loss returns").

The cost of the lost mud, at up to $500 a barrel, was relatively minor in the grand scheme of things—even when you'd lost a couple of thousand barrels of mud, as was the case with Macondo. The bigger problem was the wasted time. Lost mud required corrective measures that could throw the schedule out of whack. You might have to shut the well in and idle your drill team while you waited for more mud to arrive on a slow boat from who knows where. The stomach-churning anxiety on a rig is not that disaster will strike, but that the rig will have too much downtime, get too far behind schedule and over budget, and be deemed an inefficient operation. For oil company executives, efficiency is critical. Time is money. At a million-plus bucks a day, losing one day is worse than losing all that mud.

So the engineers on the Macondo job had to wince every time someone talked about "pore pressures" and "fracture gradients" in this dreadful hole. There was almost no margin of error. The mud had to be strong but delicate. This was baking a soufflé in the hot, dark, deep interior of the planet.

And it was even worse than that, actually, because, as the drillers discovered, the Macondo prospect was not a single, coherent, placid, ripe, eager-to-be-exploited pool of oil. No, it was an unruly gaggle of reservoirs; a complex, layered array of ancient sands, some of them filled with oil, some with briny water, each with its own geological composition, its own personality and demands and dietary requirements. There were five distinct layers of fluid-bearing sandstone. The top was a brine layer, highly pressurized, no oil—not a "pay sand," as they say. The oil and gas were largely in the bottom three layers, which had lower pressure than the top, brine layer, and could handle only lighter mud. So it was confounding: The mud weight necessary to drill through the higher layers would turn out to be too heavy and brutal for the lower layers. The mud engineers had to get out their calculators and mud-weight conversion tables, grinding through the math, converting surface pressures to pressures at depth, figuring out what the mud weight should be at each step of the way.

This well was a mess. The well team engineers in Houston fussed over their plan, agonized, and racked their brains for solutions. ("This has been a nightmare well which has everyone all over the place," engineer Brian Morel emailed a colleague.) They finally made a sound decision: They would declare victory and retreat. Having reached 18,360 feet, with more than 1,500 feet still to go on their well plan, they called it quits, declaring that they'd gone deep enough.

That didn't mean they were going to abandon Macondo permanently. This well was a "keeper": an exploratory well that had been designed to be converted to a production well. Old-school drillers didn't like these conversion jobs, because they added a layer of complexity to what was already a tricky operation. BP, however, wanted this well to be more than a diagnostic hole. It wanted Macondo to produce a revenue stream for the company. This problematic hole needed to pay off directly.

The BP engineers came up with a plan for setting the final string of casing and then cementing the well. Collaborating with the cement contractor, Halliburton, the BP engineers decided that, given the mud losses, they should use a nitrogen-foamed cement slurry. They calculated that the foamed cement would be lighter and less likely to leak into the formation, but it would be strong enough to plug the bottom of the well.

Early in the morning of Monday, April 19, Halliburton workers on the rig sent sixty barrels of cement down the well.

It looked like a good cement job. The rig crew watched to see if the amount of mud coming out was the same as the amount of cement going in, or "full returns." That's what they saw. There were other signals that it had gone well, including the lift pressure: the amount of pumping pressure required to push the cement to the bottom and then back up the space between the inner casing and the outer well bore.

On Tuesday, April 20, the day that the executives were scheduled to arrive, the well team at BP's office in Houston and the managers on the rig held their usual 7:30 a.m. conference call. They went over the next step in the well-completion plan. They had put together a "decision tree." If certain things happened, certain decisions would be made. The decision tree was clear: If the early signs on the cement job looked good, and they had full returns, they could skip the time-consuming test known as a cement bond log. That test, which would have taken at least twelve hours, would have sent an instrument down the well to take acoustic readings of the hole and search for signs of gaps in the cement. Three employees of the firm Schlumberger, who were on the rig to conduct the test, were told that, since there was no hint of problems with the cement job, their services would not be needed. They took an 11:00 a.m. flight back to land.

The Pressure Test

Jimmy Harrell, the offshore installation manager—Transocean's top employee on the rig—showed the executives around the Horizon. Harrell was popular, a good ol' boy with white, thinning hair, a walrus mustache, and a low grumble of a voice. He looked a little like the character actor Wilford Brimley. He'd had every job in the offshore industry, from floor-hand to derrick man to driller to toolpusher. He'd been with Transocean since 1979—longer even than Daun Winslow—and on the Deepwater Horizon since January 2004. He loved his men dearly and was proud to be the boss of an elite rig doing challenging work in the Gulf of Mexico. He had spent most of the morning and early afternoon of April 20 preparing for the arrival of the VIPs, figuring out where to take them, what to discuss with them, and who to introduce them to. He planned to be with the executives most of the afternoon and evening.

Also along for the tour was Captain Curt Kuchta, the rig's master. He'd had the position on the Horizon for less than two years, and was a couple of decades the junior of Harrell. Kuchta was known on the rig as "Captain Curt." When the rig was "on station," linked to the well—as it was on this particular afternoon—Jimmy Harrell had command of the whole operation, but when the rig unlatched from the well and began to move, the command would shift to Captain Curt.

Harrell took the executives down to the moon pool, the open space just above the water where the riser emerges from the deep. The men looked at the legs of the rig and discussed a maintenance problem with one of the thrusters that controlled the rig's position.

They visited the storage lockers. Pat O'Bryan of BP asked why some of the safety equipment lacked labels, and he learned that everything was now equipped with infrared radio tags.

The executives moved on. A routine visit, this.

At a quarter to five, the four visitors reached the center of the action on the rig, the drilling shack—the "doghouse." Two men sat in swivel chairs in front of a panel of monitors. The shack was air-conditioned, and the executives were happy to escape the heat momentarily. But they quickly sensed that it might not be the best time to intrude upon the operation. The doghouse was unusually crowded even before they entered. The drilling team was having an animated discussion with the BP well site leader on duty, Bob Kaluza.

Kaluza had been on the Horizon for just four days. He had replaced a well site leader who had flown to shore for routine training. The well site leader is traditionally known as the "company man." He's the top BP person on the rig. What the company man says is law, pretty much. The rig workers can protest, and they can halt operations if they sense that safety is being compromised. But there's no confusion about who's ultimately in charge: the company man.

Daun Winslow sensed that things were a bit tense in the shack.

"You got everything under control here?" he asked the driller, Dewey Revette.

"Yes, sir," Revette answered.

"Let's go, and let them guys do their work," Winslow said to the other executives. He suggested to Harrell and Randy Ezell, the senior toolpusher who had joined the tour, that the two of them stay in the doghouse to help out.

The drill team on this afternoon was displacing the heavy drilling mud from the riser and from another 3,367 feet of casing below the wellhead. That was a lot of mud, and it had to be off-loaded to a supply boat adjacent to the rig, the *Damon Bankston.* The crew had already conducted a couple of pressure tests that pressurized the well to ensure that the seals and casings were holding properly. Now they were in the process of a third pressure test, called a negative test, which diagnosed the quality of the cement job at the bottom of the well. Such tests almost always prove successful. The test "underbalances" the well by removing so much heavy mud that the pressure from above is less than the known reservoir pressure. That tempts the oil and gas to invade the well. The cement, however, should prevent such an incursion. The negative test lets the drill team know that the cement plug at the bottom of the well is an unbreachable barrier to the hydrocarbons.

But as the crewmen conducted the negative test, they saw something they didn't understand. When they opened the top of the drill pipe, a surprisingly large quantity of seawater (on the order of 23 barrels, though recollections vary) flowed out, and the pressure wouldn't fall to zero psi as it should have. They closed in the drill pipe and the pressure rose to 1,262 psi—not a good result. They opened it up, bled off pressure, closed it in, and again the pressure built. Opened it, bled it, made some adjustments in the well, closed it again, and this time the pressure built even more, to 1400 psi. It should have been zero.

Hence the tension in the doghouse.

The guys discussed the test result at length. They chewed on conjectures. It could be explained by a lot of things. Drilling isn't an exact science. Gauges can go on the blink. Someone mentioned something called the "bladder effect."

Along came the other company man, Kaluza's relief, Donald Vidrine, who most certainly did not like the 1,400 psi on the drill pipe. Vidrine and the drill team talked it over and eventually reached a decision: They'd do it all over again, but this time with a different method. They would test the pressure on the kill line, a three-inch line linking the rig and the blowout preventer. If there was pressure from the reservoir, they should see it on the kill line.

They opened the kill line and watched. Some fluid came out. The stream slowed to a drip. Then it stopped altogether.

They watched the kill line for another thirty minutes. No flow.

If hydrocarbons had entered the well, they would expect to see pressure on the kill line. They didn't. This was an excellent result. The 1,400 psi pressure reading remained on the drill pipe, and that was puzzling, but . . . but they had the good result on the kill line. Slowed to a drip, then no flow. That was just what they wanted to see.

Vidrine gave the order to proceed as originally planned: to keep taking the mud out of the well and pumping it over to the *Bankston.*

And so the mud came out, and the seawater went into the well.

The mud weighed 14.5 pounds per gallon. The seawater weighed 8.6 pounds per gallon.

Not a Simulation

After the executives left the drill shack, they looked at a few more areas of the rig and then repaired to their private quarters to freshen up before dinner. At six o'clock they gathered in the galley, ate their meal, and then headed to a conference room for a seven o'clock meeting. On the way, Winslow asked Jimmy Harrell, "Everything okay up on the rig floor?" Harrell gave him a thumbs-up.

The meeting lasted two hours. At one point, Vidrine got a call on his cell phone and had to return to the rig floor to talk about the negative test. He soon returned, then left a second time. It was a busy evening on the rig, but the VIPs focused on their agenda, which was to discuss safety.

The two BP executives had come to the rig with talking points. Among other agenda items, they wanted the Deepwater Horizon crew to pay closer attention to dropped objects, the potential for hand injuries, and slips and falls.

The talking points included:

"Difficulty performing jobs may lead workers to taking shortcuts."

"What is your rig doing for hand injury prevention this year?"

"How do you assess the slip potential when stepping up or down in irregular situations?"

The talking points were full of praise for the Horizon team:

"Genuine concern for health, safety, and environment—not just ticking the box."

"Tremendous rigor put into risk awareness, mitigation, reporting, investigations, lessons learned."

"No blame, 'can do' culture—fix the problem, learn, move on."

"Prudent risk taking—freedom to fail, no fear of second guessing."

This last talking point seems to have a note of ambiguity to it. Safety came first, always, but this remained an industrial process in which you had to keep moving forward. "Freedom to fail" could cut both ways: It could mean halting a job or keeping going. "Prudent risk taking": You could put the emphasis on "prudent" or the emphasis on "risk taking." It was a delicate balance, this elaborate system of drilling for oil in the safest and yet most efficient manner.

There were so many judgment calls.

BP had stockholders. Executive compensation was based in part on the ability to keep costs under control. An oil company knows the incremental cost of extracting a barrel of oil from the earth. It knows how much each rig costs per day, and how many days, on average, a rig requires to drill 10,000 feet of well. The meter is running. No dithering allowed. Drilling isn't for slowpokes or wimps.

David Sims, BP, testifying months later: "We're a business, and we have shareholders. Our job responsibilities are to be fiscally responsible. So every conversation, every decision has some cost factor."

Daun Winslow, Transocean, also testifying: “It is a business, and you want to make sure your machines are hauling things. If you're a truck driver, if you don't run, you don't get paid, right?”

In the meeting, they discussed the timetable for bringing the Horizon in to dry dock for much-needed maintenance. No one voiced any concern about what was happening up on the drill floor.

At 9:00 p.m., the executives went to the bridge. They took turns at a simulator, a monitor equipped with software that allowed them to practice operating the rig in a variety of conditions. The software program showed them what it would be like to take the helm of the Horizon in a weather emergency. They navigated their way through a simulated tempest.

Winslow: “We loaded into the simulator, you know, about seventy-knot winds and thirty-foot seas and two thrusters down, and then you switch it into the manual mode and see if the individuals can maintain the rig on location.”

By twenty minutes after nine, Randy Ezell, the senior toolpusher, had knocked off work and was back in his room. He called the rig floor to make sure everything was okay. He spoke to Jason Anderson, the toolpusher. Anderson was finishing the final job of his tenure on the Horizon. He was scheduled to take a helicopter flight the next morning. A big man, thirty-five years old, Anderson had been promoted, and he was looking forward to joining the crew of a rig named the Discovery Spirit, where he'd be a senior toolpusher—*senior,* up a level on the flowchart.

Ezell recalled: “I said, ‘Well, how did your negative test go?’ He said, ‘It went good.’ He said, ‘We bled it off. We watched it for thirty minutes, and we had no flow.’ And I said, ‘What about your displacement? How is it going?’ He said, ‘It's going fine. It won't be much longer, and we ought to have our spacer back.’ I said, ‘Okay. Do you need any help from me?’ He told me, ‘No, man.’ Just like he told me before, he said, ‘I have got this.’ He said, ‘Go to bed. I have got this.’ He was that confident that everything was fine. I said, ‘Okay.’ ”

Ezell called his wife and told her that things were good on the rig. He got in bed. He turned on the TV and flicked off the light.

Jimmy Harrell, back in his own room, took a shower.

Mike Williams, the chief electronics technician, called his wife.

Through the phone, she could hear an alarm going off. A gas alarm. She asked him if he needed to get off the line. He assured her it was not a big deal. There is nothing unusual about hearing a gas alarm on a drilling rig that is actively handling a gassy well.

Williams: "I'd get to the point where I didn't even hear them anymore, especially with this well because we were getting gas back continuously. It was a constant fight."

Then he heard a sound he hadn't heard before. A hiss.

"I need to check this out," he said and hung up.

Daun Winslow by this point had quit the simulator on the bridge and roamed down to the galley to get a cup of coffee. There was a small space set aside for smokers.

He lit a cigarette.

The survivors have different stories of what they were doing, what they heard, what they felt, and what they saw, but there is a common thread to the narratives: No one saw it coming. They were doing routine things with no sense of impending calamity. They didn't hear the ticking of the time bomb. The first warning, for many, was the shaking, the trembling, the deep and powerful vibration that gripped the massive rig. On this calm, serene night, with a current of only two-tenths of a knot, the Deepwater Horizon shuddered to its bones.

On the bridge, the remaining executives stopped fiddling with the simulator and looked around to see what in the world was going on. Captain Curt opened the port-side door, which looked out onto the water. He could see the *Damon Bankston,* the supply boat that had been taking mud from the rig. Something was raining on the boat. It was a black rain. Mud. Mud from the well.

Yancy Keplinger, one of the two dynamic-positioning officers, aimed a closed-circuit TV camera at the drill floor. He could see a pipe ejecting mud out the side of the derrick.

Down on the *Bankston,* Captain Alwin Landry watched mud splattering everywhere. A line must have broken, he told himself. Mud lines break when you off-load from a rig to a boat. Except this mud was mixed with chunks of cement.

"What is *this*?" Landry asked a crewman.

Then came the hiss, a sickening sound that they weren't supposed to hear.

Gas . . .

And no wind to blow it away.

The gulf was as flat as a pond . . .

The gas and condensate, emerging from the well, flowed across the drill floor, spreading everywhere. The green fog worked its way into the room that housed engine no. 3.

The engine began to race, going faster, getting louder . . . faster . . . overspeeding . . .

The rig went dark.

Five seconds later . . .

Chapter 2

Inferno

Months later, the blowout would be depicted in a highly technical, bloodless BP PowerPoint presentation, with graphics showing the gas breaching eight distinct barriers—like eight slices of Swiss cheese whose holes just happened to line up perfectly.

We'd hear of the cement slurry, nitrogen separation, the shoe track, the gas diverter. We would see where the gas flowed and where it encountered an ignition source. In a BP animation, the gas was depicted as a cobalt blue fluid, rising from the well, oozing to and fro, and slathering the rig until it finally found a spark. The BP presentation benefited from many months of interviews by a team of fifty investigators, and it depicted a freeze-frame disaster with all the vectors identified and calculated.

And there would be other investigations, and they would list the "fifteen risk-based decisions" prior to the blowout, and we would see the email traffic, and learn of the hallway conversations in Houston and disagreements on the rig floor of the Horizon. Things would get clearer and clearer as the days became weeks and the weeks became months, and there would come a point where the disaster would seem almost deterministic—an orderly, if tragic, sequence of events that were always faithful to Newtonian laws.

But on April 20, when the well blew, there was only chaos, noise, darkness, and terror, as the fire roared in the derrick and screams filled the damp gulf air.

The rig had 126 people aboard representing thousands of years, collectively, of drilling experience, but not one person had ever been through anything like this. They had trained for emergencies, but usually on Sunday mornings. Their fake emergencies had all the drama of going to church. They'd never trained with the rig coated in slippery mud vomited from the well. They'd never trained with a whole section of the rig simply obliterated, and walls and ceilings caving in, and people missing, people injured, people dazed and terrified and crying, people unable to recognize the crewmates they'd known for years, people now jumping—jumping!—off the rig, a seventy-five-foot plunge into water lit by the glow of an inferno.

There are few fears more visceral and motivating than the fear of burning to death.

Randy Ezell, the senior toolpusher who had just assured his wife by phone that all was well, and was watching TV in bed, got a call from one of the guys on the drill floor:

"We have a situation. The well is blown out. We have mud going to the crown."

Ezell asked him if they'd sealed the well down at the blowout preventer: "Do y'all have it shut in?" Ezell asked.

"Jason is shutting it in now. Randy, we need your help."

Ezell grabbed his coveralls from the hook, put on his socks, opened the door to his room, and headed toward his office, where he kept his boots and hard hat. He was at the office doorway when a blast threw him twenty feet against a wall.

He found himself buried in debris. He struggled to get up, but the debris weighed on him. He faltered. Finally the adrenaline propelled him up onto his knees, and he began crawling. The living quarters had been demolished. Something moist was in the air. Droplets, it felt like. Methane condensate, he later reckoned.

He put a hand on a body. Didn't recognize the man. The body still had life in it.

A beam of light came down the hall, someone with a flashlight, and then Jimmy Harrell, the boss, came out of his room wearing coveralls and no shoes, blinking, stuff in his eyes: fiberglass insulation.

Ezell saw a pair of feet sticking from beneath the wreckage. A voice emerged: "God help me! Somebody please help me!"

Ezell pulled the debris away and saw that it was Buddy Trahan, the Transocean executive who had come with Daun Winslow on the chopper that afternoon. Trahan had a deep cut on his leg, a mangled lower calf. His fingernails were gone, and he had a hole in his neck. His back had been burned from head to belt.

The inferno intensified.

Mike Williams: "The fire at this point is completely out of the top of the derrick. Things are popping, things are falling, things are starting to fly. There's projectiles coming from everywhere. There's just stuff flying everywhere."

Daun Winslow ran from the accommodations area and saw the derrick on fire, people shouting, people down in the water. He came across a crewman dangling above the water as he clung to the outside of a handrail.

"Hey, where you going?" Winslow said. "There's a perfectly good boat here. Do you trust me?"

Don't know you, the man answered.

At that moment, it didn't matter how high up the chain of command Winslow might be, how much money he made, how many years he'd been on various rigs in the gulf: He was a stranger backlit by fire.

Mark Hay, the senior subsea supervisor and one of the veterans on the rig, appeared by Winslow's side. Winslow said to the dangling man, "You trust *him*?"

The man said he wasn't sure. He hesitated. Gradually, Hay and Winslow coaxed him off the handrail and sent him to a lifeboat.

> Doug Brown, Chief Mechanic:
>
> The first explosion basically threw me up against the control panel that I was standing in front of, and a hole opened up underneath me, and I fell down into the hole—into the subfloor where all the cable trays and wires are located at. I was wondering what was happening. I was confused. I was hurting. I was dazed, and I proceeded to try to get up, and the second explosion happened. And I ended up falling back down in the hole, and the ceiling caved in on top of me at this point.
>
> After that I was wondering if there was going to be more explosions, and I started hearing people screaming and calling for help, that they were hurt, they needed to get out of here. So I

> proceeded to crawl out of the hole and looked around, and Mike Williams, an ET who was in the next room to us, was crawling over the rubble next to me, and he was heading toward the open hatch door into the—aft of the engine control room, which had been blown open and bent. He was dazed, confused, he was screaming he had to get out of here, he had a wound on his forehead, and he was bleeding profusely.
>
> We proceeded to go up to the main deck, where we saw the derrick on fire. We proceeded then to go along the port side, past the port crane, up to the bridge. Upon entering the bridge, it was complete chaos.

At 9:50 p.m., on the bridge of the Horizon, Andrea Fleytas and Yancy Keplinger, the two dynamic-positioning officers, saw combustible-gas alarms going off everywhere on their screens. Fleytas, one of the few women on the rig, had been stationed at the main console as Keplinger tended to the VIPs at the simulator. The alarm lights were magenta, the color code for the highest amount of gas.

Someone from the drill floor, where the crew was frantically trying to close the well, called the bridge.

"We have a well control problem."

And hung up.

The rig blacked out. Then came the first explosion. Keplinger, looking into the closed-circuit TV feed, could now see flames on the rig floor.

Someone came through the starboard door and said there were three or four people in the water.

Keplinger got on the intercom: "Fire! Fire! Fire!" Report to emergency stations, he ordered.

Fleytas hit the general alarm for the rig and sent out a Mayday call to other vessels in the area.

Steve Bertone, the chief engineer, arrived on the bridge and went to his station, the port-side computer. He saw there was no power, no engines, nothing. He called the engine control room, extension 2268, got no dial tone. Hung up, tried again. Nothing. "We have no coms!" he shouted.

Chris Pleasant, the subsea engineer, came into the bridge and said, "I'm EDS-ing"—*ee-dee-essing*. Emergency Disconnect System.

As a subsea engineer, Pleasant was one of the workers in charge of the blowout preventer, the huge stack of valves down there on the well. By hitting the EDS button—a prominent red button sitting amid a profusion of buttons and gauges on an instrument panel on the bridge—Pleasant could detach the riser from the blowout preventer, disconnecting the rig from the well. That would simultaneously activate the blowout preventer's blind shear rams, designed to slice the drill pipe, shut in the well and stop any upward flow of gas and oil.

"Calm down, we are not EDS-ing," Captain Curt Kuchta answered, according to Pleasant's later testimony.

But the rig is on fire, alarms are sounding. Pleasant went to the panel. "I'm getting off here," he said.

Donald Vidrine, the BP company man, said, "Yeah, hit the button."

Pleasant hit the button.

The captain asked Daun Winslow if it was okay to EDS.

"You haven't already?" Winslow said.

Kuchta turned to Pleasant: "You can EDS."

"I already did," Pleasant said.

Stephen Bertone: "I hollered to Chris, 'I need confirmation that we have EDS'ed.' He said, 'Yes, we've EDS'ed.' I said, 'Chris, I need confirmation again. Have we EDS'ed?' He said, 'Yes.' I said, 'Chris, I have to be certain. Have we EDS'ed?' He said, 'Yes,' and he pointed at a light in the panel."

Chaos, confusion, fear. What the hell was going on? And who was in charge? Captain Curt? Jimmy Harrell? One of the visiting executives? If so, would that be the top BP guy or the top Transocean guy?

Into the bridge rushed a man covered with blood. He was unrecognizable. He said, "We have no propulsion, we have no power. Engine number three for sure has blown up. We need to abandon ship now."

The others finally identified him: Mike Williams, gashed in the forehead. Bertone searched for a first-aid kit, found nothing, and brought Williams a roll of toilet paper and told him to hold it against his head.

Pat O'Bryan, the BP executive, saw crew members mustering at the lifeboats. He turned to his colleague David Sims and said, "We need to go."

The executives left the bridge and hurried to lifeboat number two.

It was really a capsule, shaped like a giant lozenge, with windows now coated in mud and gunk. It was nearly full.

Winslow was last to get into the lifeboat. He hesitated, one foot on the lifeboat, one still on the rig. Stay or leave?

He was the senior Transocean employee on the rig at that moment. There are certain expectations. There are codes of the sea. But he was not the captain.

Captain Curt reappeared.

"I've got other men, I'm going to the rafts," the captain said. The rafts were a backup to the lifeboats. While the lifeboats were large and enclosed, the inflatable rafts were small and open air.

Winslow decided to evacuate. He told the coxswain to lower the lifeboat.

The coxswain shouted at the shaken rig workers packed into the lifeboat: "Calm down! Calm down!"

"Maybe you should calm down," Winslow told him. "Do you know how to launch the boat?"

"Yes, sir."

The executives and scores of other rig workers dropped to the water in the mud-spattered lozenge. The cox couldn't see where he was going as the lifeboat motored away from the rig. Winslow said he was going to open the door. Don't open it, the cox said. Winslow opened it anyway and helped guide the boat toward the *Bankston.* The cox didn't let off the throttle quickly enough, and the lifeboat struck the *Bankston* hard and bounced back. Men threw ropes and Jacob's ladders from the *Bankston.* The shaken Horizon workers clambered aboard.

On the rig, the last of the survivors gathered at the life raft. They loaded an injured worker into the raft and struggled to deploy it.

Stephen Bertone: "I went to the far side of the life raft, and I heard the injured person on the gurney start hollering, 'My leg, my leg!' I also heard Andrea screaming, 'We're going to die! We're going to die!' At that point, I honestly thought that we were going to cook right there."

The captain stayed behind, on the deck, as the raft descended. Something went wrong, the raft pivoted, it tipped, and then went into free fall, and as it hit the water Andrea Fleytas went overboard. She began swimming. Bertone jumped out and began swimming, too.

Back on the deck, the captain realized there wasn't time to retrieve the raft and prepare it for another launch. They were out of time. They would be cremated momentarily. He would have to jump.

A man appeared behind him: Keplinger, the dynamic-positioning officer from the bridge. He'd wanted to be on that raft.

"What about us?" Keplinger asked the captain.

"I don't know about you, but I'm going to jump," Kuchta answered, according to Keplinger's later testimony. And Captain Curt jumped.

Keplinger looked over the edge of the platform at the water seventy-five feet below. The raft was in the way. He didn't want to land on the raft. He waited. Waited.

Jumped.

Stephen Bertone, down below, looked up at the rig, his view blocked by a billowing cloud of dark smoke that created a ceiling thirty feet above him. "I saw a person's boots and his clothing and stuff come shooting through the smoke," he recalled. "Just before he landed, I noticed that it was Curt. He landed approximately five feet from me. Within seconds, a half a second later, another pair of boots and person came flying out of the smoke, and he was approximately ten feet from me. Just before he hit the water, I noticed it was Yancy Keplinger."

Mike Williams was still on the rig, up on the helideck. He had missed his chance at leaving on the lifeboat—he'd been busy trying to activate a backup generator—and by the time the life raft was being loaded, he thought for sure it would pop or melt and the people inside would be roasted. He decided to jump. With a running start, he hurtled himself from the rig, crossing his legs in midair and hitting the water feet first.

He surfaced, his skin seemingly afire.

> I've got oil, hydraulic fluid, gasoline, diesel, whatever it is that's floating on the water is now burning my entire body. I'm now covered in this sludge. I don't know what it is. It's burning, I can't hardly breathe, but I can feel the heat from the fire underneath the vessel. At that point, I started backstroking with the one arm and one leg that would work until I remember feeling no pain, I remember feeling no heat and thinking that that was it, I had died.

He snapped to, inhaled life, and swam to the raft.

Keplinger swam to it too. The captain was already there. They had to get the raft away from the rig. The rig was blazing, the heat was threatening to melt everything. The inferno could be seen for thirty miles. As Keplinger climbed into the raft he could see the fast rescue craft—the FRC—of the *Bankston* approach. Someone on the FRC latched the boat to the raft and tried to pull it away from the rig. But then a line popped out of the water, taut. The raft was tied to the rig—trapped beneath the inferno.

Who had a knife? No one. Pocketknives were prohibited on the rig. The fire roared, the heat increasingly unbearable. Finally someone on the FRC produced a large pocketknife. The captain, swimming, took the knife and cut the line, freeing the raft.

Everyone went to the *Damon Bankston.* One by one they scrambled up Jacob's ladders and onto the *Bankston* deck.

The Coast Guard, which had gotten a call from another rig that the Horizon was in flames and people were jumping in the water, scrambled two rescue helicopters and crews out of Air Station New Orleans, and two helicopters and a plane from the Aviation Training Center in Mobile, Alabama, plus four cutters: the *Pompano, Razorbill, Cobia,* and *Pelican.* Soon rescue swimmers in frogmen suits were popping out of helicopters and down on the *Bankston.* Medics did rapid triage, and the choppers used litter baskets to lift the injured personnel to the aircraft for transport to other rigs or hospitals. Executive Buddy Trahan, with twelve broken bones, was among those evacuated. Other helos circled the flaming rig as crew members with night vision goggles scanned the calm seas for PIWs—people in the water. Private boats began showing up. One, a fishing boat, had been fishing beneath the rig when the four fishermen smelled gas and sped away just before the explosion.

The three uninjured executives, now on the *Bankston,* tried to formulate a plan. They had to figure out how to help the crew, how to process this whole situation, and all the while, they struggled to make sure they knew who was where, who was still on the rig, who had already been airlifted onto shore. They were losing track of people.

Someone took muster. The evacuees numbered 113. But there had been 126 people on the Horizon.

They took muster again. The count rose to 115.

That's where it stayed.

Jason Anderson, the toolpusher, the man who had been on duty that night as they displaced the mud, and who had told his colleague Randy Ezell, "I have got this," was among the missing.

Dewey Revette, the driller, who had said, "Yes, sir," when Daun Winslow asked him if he had everything under control with the negative test, was missing too.

Roy Wyatt Kemp, Donald Clark, and Stephen Curtis, all of them assistant drillers, were gone. So was floorhand Karl Kleppinger Jr., crane operator Aaron Dale Burkeen, tool hand Adam Weise, mud engineer Gordon Jones, and mud engineer Blair Manuel. And Shane Roshto, just twenty-two years old, a roustabout from Liberty, Mississippi, who had told his wife that the earth did not want to be drilled here. Jones and Manuel, the two mud engineers, worked for M-I Swaco; the other nine worked for Transocean.

Of the 115 personnel accounted for, 17 were injured, 3 critically. The uninjured workers stayed on the *Bankston.* Some formed a prayer circle. Only a few senior rig members and the executives had access to a satellite phone. The rest were forced to bide their time, knowing that their loved ones would soon hear the news and wonder if they had survived.

They couldn't do a thing but stare at the mesmerizing fire.

Jimmy Harrell was disconsolate. Some of his coworkers feared that he was having a heart attack. Harrell eventually talked by satellite phone with his boss in Houston, Paul Johnson, the rig manager. Harrell, nearly blind from the insulation in his eyes and barely able to hear after the concussion of the explosion, struggled to maintain his composure. Johnson tried to comfort him, but the boss finally asked the question a boss must ask in such a situation: What happened?

"I don't know, Paul. She just blew," Harrell answered. "I don't know what happened. *She just blew.*"

Chapter 3

Hot Stabs

A black cloud boiled from the rig and billowed into the night sky, illuminated by the flashes of explosions. Coast Guard helicopters circled low over the water, their searchlights skittering across flat water littered with debris. There was no sign of the eleven missing men.

The fire was a different order of fire than anything anyone had seen before. It was like hellfire: ugly, oily, demented, inexhaustible, the central geyser of flame shrouded in blacker-than-black smoke. Even the sea surface, with a new skin of oil, burned furiously. Workboats equipped with water cannons fired upon the rig. The steel hissed. The boat captains continued to douse the Horizon even as it became apparent that they were battling a volcano with water pistols. They could have dumped half the Gulf of Mexico on the rig and still not put out the fire.

Throughout the oil drilling industry along the Gulf Coast, phones rang and people dressed in the dark. They raced on empty superhighways toward office buildings lighting up at an ungodly hour. The BP engineers piled into the emergency operations center in Building 4 of the company's One Westlake Park campus in Houston's western suburbs. The Transocean employees were just down the road at the company's headquarters at Park Ten Centre. In New Orleans, officials with the federal government's Minerals Management Service, which regulates the offshore industry, hastened to work in the darkest hours of the night.

Information was sketchy. No one knew why the well had blown out. No one knew at first if the fire was fueled by material on the rig—which carried not only flammable mud but also 700,000 gallons of diesel—or by the well itself.

But surely the well couldn't be fueling the flames. The rig must have disconnected. The word from the gulf was that the subsea engineer had hit the EDS button on the bridge. The blowout preventer must have sheared the pipe.

And there were other backups, redundancies, fail-safes. The industry understood the imperative of well control and had designed blowout preventers, or BOPs, to handle every conceivable scenario. The BOP is not some trifling piece of hardware: It weighs on the order of 450 tons. It stands five stories tall. It's a stack of valves—a "tree," contained in a rectangular frame—inelegant and clunky on the outside but a wonderland of plumbing within. The blowout preventer can do so much more than simply prevent blowouts. It is an integral piece of drilling equipment that is used in day-to-day operations as the crew attempts to diagnose the condition of the well, isolate fluids, and regulate pressures. It's as versatile as a Swiss Army knife. It can do everything this side of scramble an egg.

The stack of the Horizon's BOP (including, to be precise, the lower marine riser package, or LMRP, that sat on top of the BOP proper) had different rams and preventers. Starting at the top:

Upper annular preventer
Lower annular preventer
Blind shear ram
Casing shear ram
Variable bore ram
Variable bore ram
Variable bore ram (test ram)

No wonder wells rarely explode. On any blowout preventer, there is redundancy on top of redundancy. Backups to backups. Work-arounds galore. If you can't shut in the well with the upper annular preventer, you can use the lower annular preventer, or one of the variable bore rams. And there's always that ultimate backup: the blind shear ram. The big pinchers. They're the most important of the rams. The paired blades of the blind shear ram are slightly offset from one another, closing like a powerful vise to cut the pipe and seal the well. Heavy rubber "packers" around the blades are supposed to form a tight seal, closing all flow in the well.

The engineers who designed the BOP realized that there might be situations in which, for whatever reason, the rig crew is disabled, or the rig loses power, or the rig's dynamic positioning system fails and the whole rig drifts away, and the well needs to be sealed. So there was the AMF: automatic mode function. Also known as the "dead-man switch." The engineers had thought of everything: The BOP, sensing something wrong, would take drastic action *by itself* to seal the well.

So much engineering mastery had gone into this drilling process, into the hardware, the procedures, the protocols. People had anticipated things going wrong and had designed corrective measures. And yet somehow the impossible had happened, and the Horizon floated on the gulf amid a riot of implacable fire.

At the end of this traumatic night, Daun Winslow found himself in charge of figuring out what to do next. The Coast Guard could handle the search-and-rescue operation, but it didn't fight fires. It was Transocean's rig, and Transocean's fire, and of all the Transocean people at the scene, Winslow had seniority. Winslow didn't get any official word that he was in charge, but he assumed command. A battlefield commission, this was.

Daybreak brought a stunning sight: The volcano's plume soared thousands of feet in the air, a pillar of smoke that eventually reached jetliner altitude and flattened out.

Winslow and the other top managers studied the motion of the burning rig. If detached from the well, the rig should drift with the currents. But the Horizon was inscribing a circular path on the sea surface. There was no doubt anymore: The rig hadn't disconnected. It had an anchor to the sea floor. The Horizon was tethered to Macondo by a mile-long pipe.

The flammable materials on board would have burned up by now. This had to be a well fire. The Horizon had been turned into a blowtorch, fueled by the Macondo reservoir.

The rig began to list in the water.

The Horizon survivors on the *Bankston,* the ones who hadn't been evacuated to hospitals, were stuck for hours and hours at the scene of the disaster. The Coast Guard, controlling the scene, initially would not release the *Bankston,* which had been assisting with the search-and-rescue operation. Not until very early Thursday morning, almost twenty-eight hours after the explosion, did the *Bankston* finally deliver the survivors to dry land in Port Fourchon, Louisiana. The survivors were handed plastic cups and directed toward the portable toilets. After a maritime accident, the Coast Guard requires drug tests. That accomplished, the survivors piled into company buses for a final two-hour drive to the Radisson Hotel near the New Orleans airport, where they were given rooms and, finally, a chance to reunite with their families.

But in the first hours of the disaster, the families of the rig workers struggled to obtain reliable information about the explosion offshore. At seven in the morning on April 21, Steve Gordon, a maritime injury attorney in Houston, got a phone call from a woman named Tracy Kleppinger. She said her husband, Karl, worked on the Horizon and that she hadn't heard from him or from the company. She'd read about Mr. Gordon online and knew that he'd handled a recent wrongful-death case involving Transocean. Please help me find my husband, she said.

Gordon still had Transocean contact numbers in his cell phone. He got bounced around, gleaning little information. But at about two in the afternoon, a Transocean human resources manager called Gordon to say that Karl Kleppinger was among the eleven missing.

Gordon called his new client and tried to break the grim news. She wouldn't hear it.

"They found him!" she said. Gordon said no, that wasn't right. But Tracy insisted: There was a breaking report on an online news site that said the eleven men had been found in a life capsule. It was attributed to the Plaquemines Parish sheriff's department.

With Mrs. Kleppinger on the line, Gordon called back his Transocean contact and said that there must be some mistake, the men have been found. "Is this the Plaquemines Parish sheriff's report?" the Transocean manager said.

She said it was.

"We've traced that down and it's not true," he said.

Then came silence over the phone line—the silence of human grief, which in the hours ahead would rip through ten other families scattered across the Gulf Coast.

* * *

The people in the Houston war rooms at BP's and Transocean's offices had no hard data to go on as they tried to figure out what to do. They had drips of information laden with rumor and conjecture and the words of rattled men who'd just escaped the hellfire. What had happened? *It just blew.* The state of the well remained a mystery. Macondo was a thing out of sight, inscrutable, existing in the netherworld of the deep sea. If the well had blown out and was now fueling the fire, the engineers would have to find a way to activate the blowout preventer, even as they worried that this could backfire. A blown-out well, if suddenly sealed, might undergo such an intense buildup of pressure that it would blow out again—an "underground blowout." The hydrocarbons could blast laterally through the steel casing of the well, surge into the rock formations, and eventually migrate up through the floor of the gulf in a free-for-all of oil plumes—an oil-leak jamboree. You could conceivably empty the entire reservoir that way.

Despite the uncertainties, the engineers decided to try to shut in the well using the only technique at their disposal: They would "splash" an ROV (remotely operated vehicle).

The ROV would descend to the seafloor, latch onto the blowout preventer, and use a robotic arm to stick a hydraulic line, or "hot stab," into a port on the blowout preventer. An ROV using a hot stab was a bit like a 1940s telephone switchboard operator plugging a phone line into a socket to connect a call. The ROV would pump fluid through the line and thereby activate one of the rams in the blowout preventer, which would seal the well.

Daun Winslow would supervise the operation. He relocated from the *Bankston* to a work boat named the *Max Chouest,* which had an ROV.

They would also need the right tools. Specifically, they needed a certain kind of hot stab. The people in Houston told them that the Horizon BOP required a 17-D hot stab. But the *Max Chouest* didn't have a 17-D hot stab.

It had a 17-H hot stab.

Winslow put out a hot-stab-wanted alert to the boats and rigs in the area. They all had hot stabs, but they all had the 17-H. For a procedure on a blowout preventer that specifically requires a 17-D hot stab, a 17-H hot stab is no more useful than a turkey baster.

More than sixty years after the birth of the offshore industry in the Gulf of Mexico, this remained the realm of the customized tool, the makeshift instrument, the jerry-built this and the jury-rigged that. No one had ever created an interchangeable array of universally applicable subsea hardware. It was Mac versus PC at the bottom of the ocean.

And here's where, at the moment of crisis, a more general truth dawned on anyone who might have a spare moment to ponder it: They weren't ready for this.

They were at sea in this crisis in so many ways. People had been drilling oil wells for 151 years, since Colonel Edwin Drake struck oil sixty-nine feet below the surface near Titusville, Pennsylvania, and they had endured every kind of blowout, gusher, fire, explosion, cratering, and so on. But no one had ever faced this precise problem: a blowout in deep water. They didn't have the tools to fix it.

What they really needed at this moment, as they hoped to fiddle with the Deepwater Horizon's blowout preventer, was the Deepwater Horizon's ROV. The rig's ROV had all the tools to prod and poke the rig's blowout preventer. The industry viewed that ROV as one of the redundancies on the blowout preventer: If the blowout preventer failed to seal the well, the rig crew could splash the ROV and manually activate the rams. Except this redundancy was now caught up in the initial calamity: The Horizon ROV at this point was very likely a hunk of melted steel.

After a prolonged search, Winslow and his comrades tracked down a 17-D hot stab on a nearby drilling rig, the Deepwater Nautilus. The next step was to deploy ("jump" is another industry term) the *Max Chouest*'s ROV. But to do that, they needed to move the boat closer to the inferno. The limiting factor on an ROV is its tether. The submersible, parked inside a cage that functions as a kind of garage, is first lowered from a cable. When the submersible reaches the seafloor, it leaves its garage and roams on a tether. The *Max Chouest* needed to maneuver within 1,000 feet of the spot directly above the well so that the ROV could reach the blowout preventer.

Back in Houston, engineers debated whether the *Max Chouest,* or any other ROV boat, would get roasted if it approached so close to a rig fire. Maybe they could run a computer model. Maybe they could calculate how close to a fire of X intensity a boat of Y dimensions and Z fire-retardancy could approach before it ran the risk of becoming burning vessel number two.

Absurd, thought Doug Martin, president of the Smit salvage company, who had raced to Transocean's headquarters early that morning. He thought: a model at a time like this! No one had any data. How hot was the fire? Was it growing or diminishing? Models are useless without meaningful inputs. Garbage in, garbage out—that would be your computer model, he figured.

At the scene, Winslow ordered the workboats with the water cannons to create a curtain of water to keep the *Max Chouest* safe as it maneuvered into position. The Horizon had drifted about 500 feet off station by that point. The *Max Chouest* eased within 400 feet of the well, upwind and upcurrent.

Late in the afternoon, the *Max Chouest*'s ROV descended into the murk of the Gulf of Mexico. The trip to the bottom took ninety minutes.

It was a long way down.

The Submersibles

The mud hole has its subtle charms. The seafloor is not always a flat and undifferentiated expanse of mud, but rather has ridges, knolls, canyons. At the seafloor, the "mud line," there's silt, mostly. The mud in spots can be hundreds of feet deep. In some places, boulders of crystallized methane, called methane hydrate, protrude from the mud line.

The very deep sea is not lifeless—*azootic*—as scientists believed well into the twentieth century. Rather it is lousy with crabs, sponges, sea anemones, octopuses, jellyfish. Even in the lightless realms, 3,000 feet below the surface, or even fully a mile down, there are coral reefs—Lophelia coral, it is called. Coral gardens grow on the wrecks of freighters sunk by German U-boats in 1942 and 1943. Where photosynthesis fails, chemosynthesis fills the void. Tube worms flourish where methane gas and oil seep from the seafloor.

For some bacteria, oil is food. Other bacteria metabolize methane gas. Life on earth, astonishing in its adaptability, in its jazzy improvisations, makes the engineering masterpieces of human beings look clunky and primitive.

In addition to the tiny organisms, there are also megafauna in the deep. Sperm whales have formed a colony near the mouth of the Mississippi; they dive thousands of feet into the darkness to feed on giant squid. Whales have been known to rub themselves on oil well risers in the same way that grizzly bears scratch themselves on pine trees.

There are eels down there. A jumbo eel, in fact, would soon take a close look at the Horizon's blowout preventer—a moment captured on underwater video.

The ROV pilots say that, once upon a time at Shell's Perdido field, which at 8,000 feet is the deepest operating well in the gulf, a giant squid attacked an ROV. There is video on the Internet, shot by an ROV at Perdido, that shows a creature barely known to science, the big-fin squid *Magnapinna.* Rather than two long tentacles and eight short ones, which is the normal arrangement for cephalopods, *Magnapinna* has ten identical arms that hang from sharply jutting elbows the way marionettes hang from a puppeteer's sticks. The squid's limbs drag the seafloor in a trawling maneuver.

The pilots have also seen the ugliest creature of the deep: the giant isopod *Bathynomus giganteus,* "the sea cockroach." It's the size of an armadillo, patrolling the seafloor on many scuttling legs, with two wiggling antennae jutting from a pinched, space-alien face. It scavenges for dead and rotting flesh. ("Think of a giant roach, put it on steroids," says Thomas Shirley, a marine biologist at Texas A&M University.)

The water is less murky in the very deep water. The oxygen level drops, and there aren't as many bacteria. Life adapts to all conditions, but at great depth it is a thin stew. The water clarity comes with a caveat, which is that you can't see a thing without artificial illumination. Save for the twinkle of some bioluminescent plankton, there's nothing but darkness down there at 5,000 feet.

Until the ROVs show up. Remotely operated vehicles make deepwater oil drilling possible. They're the spaceships of the deep.

The world at depth, viewed through the ROV cameras, is nearly monochromatic. The water absorbs the red end of the spectrum. Only when viewed up close, and brightly illuminated, will an object reveal the full range of its color. Otherwise this world is uniformly blue-green, the colors as flat as the video screen.

There is permanent marine snow in the deep sea, a cloud of dead organic matter drifting down from the top of the water column. It will float through the visual field, seemingly weightless, which adds to the sense that this is a scene from outer space.

The water is a natural magnifier, so the pilots must learn to judge the correct distance to an object. They can't tell how big something is, because there's no landscape reference point. A 50-foot tall blowout preventer can look no larger than a refrigerator.

The ROV pilot lacks a sense of touch as he manipulates the craft that is 5,000 feet below him. The pilot can watch the robotic arm make contact with the subsea hardware, but he has no texture to analyze, and is forced to rely entirely on visual cues to understand if something is hard or soft or wiggly or slippery, if it's momentarily recalcitrant or completely stuck beyond any hope of budging.

The early ROVs, back in the 1980s, had just one arm and had trouble staying in position. The second arm was added primarily to serve as a grabber. Back then the video was poor, like watching a ballgame on a 1950s television console while Dad screamed out the window at the poor kid on the roof who was assigned to adjust the aerial. The cameras weren't as sophisticated. And the technology was so new that no one really had become good at flying the vehicle or manipulating the arms. The pilots were akin to barnstormers, and if there had been barns in the deep, they would have crashed right into them.

As with so many technological breakthroughs, the birth of the ROV owed much to the military. The earliest unmanned submersibles were used for torpedo recovery. The breakout moment for ROVs came in 1966, after a US Air Force B-52 bomber collided with a refueling tanker high over the Mediterranean Sea, dropping four hydrogen bombs. Three of the nukes were recovered on land, but one went into the sea. After a manned submersible failed to recover the warhead, an unmanned sub known as CURV (for cable-controlled underwater research vehicle)—designed to pluck torpedoes from the seafloor—retrieved the H-bomb.

In the 1970s the telecommunications industry began using ROVs to examine undersea cables. In 1973 an ROV raced to the rescue of two men trapped in the *Pisces III,* a manned submersible that had been deploying transatlantic cable and, suffering a flooded buoyancy tank, had sunk to the seafloor at a depth of 1,575 feet. The ROV helped bring the *Pisces III* back to the surface after the men had endured a seventy-six-hour ordeal and were on almost literally their last gasp. Among other highlight-reel moments in recent years, an ROV in 1999 retrieved Mercury astronaut Gus Grissom's *Liberty 7* space capsule from the bottom of the Atlantic. (The sinking of the spacecraft in 1961, and Grissom's near drowning—the astronaut vehemently denied accusations that he had panicked and popped open the hatch too soon—is a case study among academics in how even the most carefully planned system is vulnerable to catastrophic failure.)

But the ROV conquest of the oil industry took a long time. Through the 1970s and into the 1980s, human divers still dominated the business. They would dive as far as 1,000 feet, enclosed in specialized chambers, enduring protracted, dangerous deployments that required up to ten days of decompression afterward. It was a costly operation that could require a thirty-person backup crew.

Gradually the machines improved. As the offshore industry started going into the deep water, the people who paid the bills had the same revelation that many NASA officials have had over the years: It's cheaper and safer to do this work with machines, and beyond about 1,000 feet, it's the only way.

The oil industry's subs were tiny at first. An operator could carry an ROV in his arms, as if it were a baby. They were basically eyeballs on tethers.

Some have called ROVs "robotic submarines." A top executive at BP, Lamar McKay, used the term "robot-controlled submarines" during the early days of the Deepwater Horizon disaster, when the American people were just figuring out what was happening and what the terms were, and how the oil industry does this crazy thing of drilling holes in the bottom of the Gulf of Mexico—but *robotic* is a misnomer. An ROV is simply a machine that, as its name says quite explicitly, is operated remotely. There's little automation. There's no computer programming that tells the ROV how to behave in situation A or situation B. The ROV can't decide on its own whether to go left or right, or open a valve, or snare a buried cable from the mud floor of the gulf. It's no more of a robot than a radio-controlled model boat puttering around a pond in Central Park. An ROV is an elaborate and expensive power tool. It is always tethered to a surface ship, receiving commands on a hard line. It's about the size of a small car. It has lights, thrusters, and two arms that look like prosthetic limbs.

For precision jobs requiring fine motor skills, they use the right arm—the fine manipulator arm—which can turn a screwdriver and twist a hand wrench. For brute-force work, they use the left arm, known as the "Conan." "Because it's a barbaric arm; it's heavy-duty," says Tim Weiss, an ROV pilot for Oceaneering International. The Conan is designed to bust, smash, ram, pound, grab.

The ROV guys do not "drive" or "guide" or "supervise" or "steer" the ROVs. They "fly" them. They're pilots.

Charles Harwell of Oceaneering: "We have a joystick, like the space shuttle and a fighter plane."

David Mahi, his colleague: "We fly them in three dimensions, not two dimensions."

This requires a mental trick, a changing of dimensional gear as they take that flat image on the video monitor and translate it for the 3-D reality of the subsea environment. There is no school for this. They master the craft through what everyone in the offshore trade calls OJT, for on-the-job training.

There are currents in the deep. The mud hole is not an inert and predictable pond, but rather is a sea of significant size, filled with currents large and small. Part of the pilot's job is staying in one place. Grab something, hold on, keep it steady. Anticipate where you'll be in five seconds, ten seconds—dimension four kicking in. The pilots struggle to maintain elevation, using thrusters to overcome their craft's natural buoyancy (a design feature to allow the ROVs to rise to the surface if disabled). Thrusting down is better than thrusting up, at least when near the seafloor, because thrusting up will produce a silt storm that can take forever to clear.

The guys sit in comfortable pilot chairs. Invariably, people will tell them that what they do sounds a lot like playing a video game. But the stakes are different. It takes days or even weeks to create the "storyboard" of a subsea operation. There's an elaborate procedural narrative that has been scripted by engineers on land, and the ROV guys, out on the water, have to act out their role, and do it with a machine that might be a mile below them, puttering through darkness, fighting currents. Most of all, the pilots must not break the expensive machinery that the oil company has spent millions of dollars placing carefully on the seafloor.

"This is the kind of video game where you can't put in another quarter and get another chance," Tim Weiss says.

The Hot Stabs

David Hayes, deputy secretary of the interior, arrived at his office that Wednesday morning, April 21, with no inkling that it would be anything other than a normal workday. He heard about the explosion within minutes of walking into Interior's massive building a few blocks from the White House. He had no concept of the seriousness of the situation. He learned that the search-and-rescue operation was under way. The fire could be contained, presumably. But by midmorning the scale of the disaster became more apparent, and Hayes's boss, Interior Secretary Ken Salazar, suggested that he jump on a plane and fly to New Orleans.

Hayes would rather have stayed put; it was his daughter Molly's eighteenth birthday. He called his wife: "We may have a problem here."

He called Admiral Thad Allen, the Coast Guard commandant, and learned that there was an early afternoon news conference in New Orleans. Allen would push the meeting back to accommodate Hayes if he could get to the gulf quickly. Hayes and press secretary Kendra Barkoff raced to Reagan National Airport—no luggage, not even toothbrushes—and talked their way onto a US Airways jet that had already closed its door and was about to taxi toward the runway.

They made the afternoon press conference at Coast Guard headquarters, then raced to the Minerals Management Service office in the suburbs. Officials had set up a monitor with an astonishing video feed from the gulf. It was live imagery from the seafloor, taken by an ROV. They were looking at the blowout preventer. "I had never seen ROV footage. It was remarkable to watch the ROVs work on what looked like a moonscape—5,000 feet down. It was like Apollo 13," he said.

The camera was trained on a pressure gauge. The officials explained to Hayes what was happening: The ROV was pumping fluid into the blowout preventer to activate a ram that would close the well.

Except the gauge didn't move. Something wasn't working. Hayes was witnessing, in real time, with this futuristic imagery from the deep sea, the paramount rule of Macondo: Nothing goes right.

Engineers in Houston tried to email Daun Winslow instructions for using hot stabs on the Horizon blowout preventer. The procedures were contained in large digital files. Winslow could get a simple email message on a laptop computer aboard the *Max Chouest,* but the colossal digital files overwhelmed the feeble wireless connection.

Houston came up with a work-around: The engineers broke the instructions into smaller digital files and sent them again. Still too large.

Meanwhile, Doug Martin, the salvage company president, had grown worried about the extraordinary amount of water being dumped on the rig. The Horizon and rigs like it are designed to prevent liquids from wells from spilling off the deck into the sea. The deck fluids typically run off into pits, basins, and tanks. Now there were five boats launching seawater onto the rig, and Martin feared that "downflooding" could sink the Horizon.

Martin tried to pass the warning to the Coast Guard, but couldn't get a good contact number. Neither could Transocean's Robert McKechnie, a manager in charge of maintenance who suddenly found himself in the thick of the response. McKechnie finally reached the Coast Guard. We're not in charge of firefighting, the Coast Guard said. McKechnie eventually got Winslow on the satellite phone. Cut back on the water cannons, McKechnie told him. Winslow passed along the instruction.

But the rig's list grew more pronounced. Increasingly it looked like a battleship that had been torpedoed. Wednesday afternoon, the derrick and the pipe racks—enormous structures that had towered over the drill floor—collapsed on the starboard side of the rig and dangled over the water.

* * *

When the first hot stab didn't work, the technicians pulled the ROV back to the surface and checked its equipment. Everything looked fine. They sent it back to the seafloor, restabbed, pumped more fluid, looked for pressure to build. Still no success.

It was, in fact, a bit like the email problem: They didn't have enough flow. The ROV could pump about three gallons of hydraulic fluid a minute. But to activate a pipe ram, they would have needed more like twenty-three to twenty-five gallons pumped in just fifteen seconds. With this weak ROV and its lame pump, they would never close the pipe ram.

Another ROV ship, the *C-Express,* arrived on site and splashed its submersible, and soon it was pumping away at the blowout preventer as well. Nothing doing, still.

At nine o'clock Winslow and his team changed strategy. The ROV used a grinder to cut through what is known as the "shear trigger pin," which was designed to activate what's called the auto-shear mechanism. But the grinder wouldn't cut the pin. They had the wrong tool for the job.

At 2:45 a.m. on Thursday, Winslow and his colleagues tried to simulate the dead-man scenario. They used the ROV to snip the cables and lines between the BOP and the rig. That should have triggered the blind shear ram. But nothing happened, again.

At 4:40 a.m., they once again tried to cut the auto-shear pin. Nothing doing.

At 7:36 a.m., finally, they succeeded in cutting the auto-shear pin. And something happened: The blind shear ram closed! Winslow was staring at the monitor in the ROV control room. "I had been up about forty-nine hours or fifty-two hours and forty pots of coffee, but it appeared to me that something happened," he later testified.

And yet this technological success translated into no visible change in the fire. This was a mystery. The pinchers had chomped on that pipe, but they hadn't sealed the well.

And so the rig burned.

And listed further.

And then . . . capsized.

The pontoons protruded from the surface amid a lake of burning oil. The thrusters meant for precision navigation now poked into the air like broken house fans.

The battle had been lost.

At 10:22 a.m., the great rig, the jewel of the fleet, slipped beneath the surface of the gulf, and dropped through the water column, bubbling and frothing, into the darkness, until it crashed into the mud 5,000 feet below the surface. It was nearly upside down.

The flames at the surface continued, but gradually diminished and went out. Daun Winslow turned to Ramsey Richards, a rig manager who had been helping out, and said, "I'm tired. I'm going to lay down for a while."

At 11:30 a.m. the next morning, Friday the twenty-third, Winslow boarded a helicopter and headed home. His twenty-four-hour "management visibility" tour of the Horizon had lasted sixty-nine hours.

The Horizon was gone, but there was something in its place: an oil slick, a nasty stain on the Gulf of Mexico, marking the site of a historic tragedy.

It was growing quickly.

Chapter 4

Crisis

David Hayes, the deputy interior secretary, flew back to Washington on Friday morning. *It's over,* he thought. The oil industry veterans had told him that the sinking of the rig would almost surely activate the dead-man switch and slam the pinchers shut. The well would be shut in. This had been a tragedy, a shocking event, but it seemed to Hayes to be finite in its extent.

When he landed in Washington, he got a call from New Orleans with news that would change his summer.

In these early days, everyone was flying blind, and so were the ROVs, with the typically clear water of the deep gulf turned opaque by the massive silt storm kicked up when the Horizon hit into the seafloor. The ROVs were kept on a short tether as they prodded the blowout preventer. They had not yet made a reconnaissance of the area. They hadn't found the rig or surveyed the riser.

The riser, the ROVs discovered when they finally circled the blowout preventer, was still attached to the well, and bent into a 5,000-foot pretzel. About five feet above the top of the blowout preventer, the riser veered sharply from vertical to horizontal, in what the engineers would come to refer to as the kink in the riser. Then it disappeared enigmatically into the murk.

The first news reports about the possibility of an oil spill were contradictory and confusing. A Coast Guard petty officer initially told reporters that the well could be leaking 8,000 barrels of oil a day, but the Coast Guard then backtracked. On Friday the twenty-third, at a news briefing at Coast Guard district headquarters in New Orleans, the federal on-scene coordinator for the Coast Guard, Rear Admiral Mary Landry, said no leak had been found.

"There is no crude oil at this time leaking from the wellhead. There is no crude oil leaking from the riser," she said in a TV interview.

The news media ran with the breaking development that the search for the eleven missing men had been called off. Their bodies were never recovered.

As the weekend approached, this appeared to be a tragic story of limited scope.

But if there wasn't a leaking well . . . why was there a growing oil slick?

An ROV finally toured the scene on the gulf floor—surveying by sonar, since the visual images were cloudy. The ROV flew along the riser, away from the blowout preventer. At one point, the riser bolted 1,500 feet off the seafloor before diving back down and reversing direction, as if to return to the well. For the ROV, this was like riding a roller coaster, one of the old-fashioned types you'd see at Coney Island.

The ROV detected something in the distance: a shadowy presence on the sonar. It was a leak. It came from the end of the drill pipe, jutting almost perpendicularly from the riser. It wasn't a very big leak.

The ROV pressed onward. Eventually it reached the end of the riser, and there, unmistakable, was the thing no one wanted to see.

The gusher featured a mix of oil, gas, sediment, and water. It was a powerful hydrocarbon river, surging into the cold dark deep water of the gulf.

Oil spill. Confirmed.

The ROVs also located the rig, about 1,500 feet from the blowout preventer.

BP passed along the discoveries to the Coast Guard. Saturday afternoon, the twenty-fourth, Landry held a news conference at a former Shell training facility in Robert, Louisiana. She was joined by Doug Suttles, BP's chief operating officer for exploration and production. Landry and Suttles in the days ahead would develop a certain division of labor: Landry would talk primarily about the response at sea and along the coast, and Suttles would focus on the subsea intervention, the blowout preventer, the ROVs, and so on.

A 1960s-vintage oil-spill law called for a "Unified Area Command" in a situation like this, with local, state and federal authorities joining forces with the private sector. Known as the National Contingency Plan, the law had been updated by the Oil Pollution Act of 1990, which was passed in the wake of the March 1989 *Exxon Valdez* spill in Prince William Sound, Alaska. The National Contingency Plan required the polluter to clean up its mess, with the federal government as a kind of supervisor of operations. It had always worked pretty smoothly, and no one on April 24 suspected that this spill would be qualitatively different from any of the many smaller spills that had happened in recent years.

On this Saturday afternoon, Landry announced the discovery that the well was, indeed, leaking. She said that federal scientists at the National Oceanic and Atmospheric Administration (NOAA) had estimated the leak at 1,000 barrels a day. She did not reveal how this had been determined. The figure made this a modest-sized spill, hardly a major threat to land. Skimmers were already patrolling the waters and collecting oily water.

The briefers didn't seem worried. Never mind panicking; they weren't even breaking a sweat. They looked competent and in control and were taking measures to deal with the situation as presently ascertained. Rigs were coming to help. There were protocols for these situations. Experts were dealing with this—and, the public was informed, they were the best experts, the world's smartest engineers.

Suttles was himself an engineer, a third-generation denizen of the oil patch. He looked a little like a young George W. Bush and spoke like an astronaut. He was articulate about hardware and procedures. He would give enough details to satisfy the news media in the short run, which is the only run the news media cares about, for the most part.

Landry had a knack for the upbeat take on a bad situation. At moments she exuded a kind of bureaucratic triumphalism. She spoke of the "tremendous work" that had been done in devising a procedure for responding to oil spills. On Sunday the twenty-fifth, with the black plume having been discovered only thirty-six hours earlier, and all efforts to activate the blowout preventer having failed, Landry highlighted the virtues of the federal oil spill contingency plan:

"What you see going on right now is an example of the progress we have made in organizing how we respond to spills. It's a comprehensive legislative framework. It's a very comprehensive response posture. As a member of the Coast Guard who has dealt with that for years, I am very thankful that we have that. It's a great system and we should be very proud of that."

The Gulf of Mexico was turning black, but the *comprehensive response posture* was coming together splendidly.

That Sunday, Landry offered good news about the efforts to contain the spill:

"In spite of the bad weather, we were able to apply an additional 6,400 gallons of dispersants to the spill area. And the dispersants are working very effectively in this region. To date we have successfully recovered about 48,000 gallons of oily water mix by the surface skimmers. I want to emphasize that a thorough investigation is under way . . . We continue to work diligently, as I said, to secure the source of the spill . . . We are very forward leaning. We have a third of the world's dispersants in this region."

A reporter asked the obvious question: Was this going to turn into another *Exxon Valdez* type situation? No, Landry said. She cited the 1,000-barrel flow-rate number.

"Let's differentiate between the thousand barrels per day and what the well, if it were completely open, would be releasing. That's the serious incident. That's the really serious spill response that would be necessary. And we don't have that situation right now. We don't have the well fully open. We have a thousand barrels a day emanating from the damaged riser and drill pipe."

Suttles warned that the efforts to clamp the leak might not work. He emphasized the depth of the well. He reminded reporters that the situation was unprecedented, that this was 5,000 feet of water. But even as he spoke, he radiated an engineer's confidence. He said there were multiple subsea response plans being developed in parallel. There were relief wells that could plug Macondo with cement at its base in the rock formation.

"We have the worlds' best experts working on this twenty-four hours a day," he said on Monday the twenty-sixth.

Landry backed him up: "I compliment BP on the fact that they're not—they're pulling out all stops. There are robust resources in place."

* * *

The well had exploded, eleven people had been killed, many more had been injured, the rig had burned out of control and sunk to the bottom, and now it was known, as a fact, that there was an oil spill from an uncapped gusher in 5,000 feet of water, the kind of blowout the industry had never before faced. And yet, for most Americans, the real *oh-shit* moment was still many days away.

People still didn't know what they were looking at. The essential nature of the event eluded most of us in the national media, too. We did not realize that the fire was not the *cause* of the catastrophe but a *symptom* of it. This was a fire fueled by a raging well that had blown out. There were some oil industry journalists on the story, but most of the reporters who covered the disaster could not have said, when this began, whether a blowout preventer is bigger or smaller than an automobile carburetor.

The national spotlight is a quirky instrument, prone to getting stuck on one thing and failing to pivot to another. The spotlight is not well designed for the scrutiny of a novel, unscripted event. We had no cultural memory of a deep-sea blowout. We were like island people who had felt an earthquake for the first time in generations and did not understand that the receding of the sea was the harbinger of a terrible wave approaching from beyond the horizon.

Maybe location was one reason that some news organizations were slow to get cracking on Deepwater Horizon as a major national story: The explosion happened offshore, in the vast Nowheresville of the Gulf of Mexico. The news media had limited ability to cover the disaster. You could fly into New Orleans and rent a car and gun it down Louisiana Highway 23 to Venice, Louisiana, where the Coast Guard had set up an emergency operations facility in a marina parking lot—but you couldn't actually see the disaster site. You could see only fishing boats and water and swamp and the occasional alligator nosing about. You might try to get on a boat and head out toward the site, but you'd be turned back, because this was now a controlled space, patrolled by the Coast Guard.

The Deepwater Horizon fire was initially overshadowed by the deaths of twenty-nine coal miners in West Virginia two weeks earlier. The Upper Big Branch mine disaster had been a dominant story, complete with a villainous coal-mine executive all but giving the finger to the government and the national press corps.

The Deepwater Horizon story broke too late at night on the twentieth for the morning papers of the twenty-first, but even on the morning of April 22, there wasn't a tremendous amount of coverage. It was, after all, the fortieth anniversary of Earth Day, and people were feeling very green. Google's logo on its home page was green. President Obama had prepared a Rose Garden message about Earth Day. His remarks included no mention of the gulf disaster. The fact that a major oil rig had gone to the bottom of the sea on the fortieth anniversary of Earth Day, which had been founded in 1970 in part as a response to the 1969 Santa Barbara oil spill, didn't generate a lot of comment initially.

The mood on Earth Day was celebratory. To a degree we may not want to admit, we spend our lives following scripts that have not necessarily been fully vetted. It's hard, if not impossible, to hit the brakes when you're going emotionally, intellectually, rhetorically in one direction at a high rate of speed. And so Obama offered nothing but happy thoughts and aspirational goals, with an obligatory pitch for Democratic energy legislation:

> Earth Day has become much more than a date on the calendar. It's come to represent the simple truth that with each challenge comes the opportunity to make the world a better place.
>
> So since taking office, we have seized that opportunity. With your help, we've made a historic investment in clean energy that will not only create the jobs of tomorrow but will also lay the foundation for long-term economic growth. We've continued to invest in innovators and entrepreneurs who want to unleash the next wave of clean energy. We've strengthened our investment in our most precious resources: the air we breathe, the water we drink, and the parks and public spaces that we enjoy.

The president had given a major speech on energy March 31 in which he signaled support for additional offshore drilling. For Obama, this was a political lever he could use to gain backing for a clean-energy bill. By expressing a willingness to allow the industry to drill some exploratory wells in areas previously off limits, including in the eastern Gulf of Mexico and in portions of the Atlantic, the Obama administration had secured the support of a key Republican, South Carolina's Lindsay Graham, on a bill that would include a cap-and-trade scheme to put a price on carbon emissions. As *The Washington Post* later documented, the administration took for granted the fundamental safety of offshore drilling when it developed its new energy policy.

The administration viewed the question through an environmental and political framework. The technology itself presented no issues. The oil companies had been drilling successfully for years, and past performance was a good, if imperfect, predictor of future success. As Interior Secretary Ken Salazar later explained, "The essence of the information I was given was that there were 40,000-plus oil and gas wells drilled in the Gulf of Mexico, and the record is the empirical basis on which you can conclude that it is safe."

But there weren't 40,000 wells as deep as Macondo, or even 4,000 wells as deep as Macondo. There were only 2,089 wells in water 1,000 feet or deeper, and only 410 of those wells were as deep as Macondo—which resided at the edge of what is bureaucratically considered "ultradeep" water.

This was never routine. This was an enterprise on the energy frontier.

"A deepwater well," deepwater oil exploration expert Richard Sears would say months later at one of the hearings of the presidential oil spill commission, "is a complex engineered system embedded in a very complex natural system."

The Deepwater Horizon disaster did generate some headlines in the early days, though nothing like what was to come down the road. *The Wall Street Journal* ran a front-page story, top strip, on April 22: "Blast Jolts Oil World." The article, reflecting the *Journal*'s business focus, did not emphasize environmental issues but instead described the possible effect of the rig fire on what had been the brightening prospects of the offshore drilling industry:

"The accident comes at a sensitive time politically for the industry. President Obama late last month proposed allowing drilling in new areas of the eastern Gulf of Mexico and off the southern Atlantic coast, where it was banned. Supporters argue the industry has become much safer. The Transocean fire could be an untimely reminder of risks."

At this point, the event still carried the Transocean brand rather than the BP brand.

Hours after the rig sank, the story was still a small headline ("Oil Rig Sinks in Gulf: 11 Workers Still Missing") on the headline-festooned home page of the Drudge Report, which normally has an exquisite eye for calamity and drama. The event did, however, make the front page of *The New York Times,* with the prescient headline "Oil Rig Sinks, Raising Fears of Major Spill."

On April 23, before the gusher had been discovered, the president's press secretary, Robert Gibbs, downplayed the incident: "I doubt this is the first accident that has happened, and I doubt it will be the last." He said Obama continued to believe in the need for expanded domestic production of oil and gas.

BP, meanwhile, seemed to be slow to recognize that that this was going to be BP's mess far more than it would be Transocean's. In its public statements, the company did nothing to claim ownership of the crisis. From April 21: "BP Offers Full Support to Transocean After Drilling Rig Fire."

> BP today offered its full support to drilling contractor Transocean Ltd. and its employees after fire caused Transocean's semisubmersible drilling rig Deepwater Horizon to be evacuated overnight, saying it stood ready to assist in any way in responding to the incident.
>
> Group Chief Executive Tony Hayward said: "Our concern and thoughts are with the rig personnel and their families. We are also very focused on providing every possible assistance in the effort to deal with the consequences of the incident."
>
> BP, which operates the license on which Transocean's rig was drilling an exploration well, said it was working closely with Transocean and the US Coast Guard, which is leading the emergency response, and had been offering its help—including logistical support.

From the statement, one might never know that there were any BP personnel on the rig, or that BP had anything to do, directly, with the blowout.

Even a couple of weeks into the disaster, BP CEO Tony Hayward would be trying to frame the spill as a Transocean incident: "It wasn't our accident, but we are absolutely responsible for the oil, for cleaning it up, and that's what we intend to do," he said on the *Today* show.

Hayward was right that, legally, BP was responsible for the spill. But BP was also responsible, as investigators would come to learn, for many of the key decisions that could have played a role in the accident. BP had designed the well and devised the strategy—with government regulatory approval—for temporarily abandoning the Macondo well. And the BP company men on the rig had responsibility for interpreting the pressure tests on April 20.

A football metaphor keeps coming to mind: A professional football coach calls a pass play in which a receiver runs a pattern across the middle of the field where a well-known bone-crushing opponent lurks at the safety position. The pass is perfectly thrown and the catch is made, but a fraction of a second later, the receiver is hit so hard that his helmet flies off and he slams to the turf unconscious. At the press conference after the game, the coach says the player was injured because the helmet malfunctioned.

There would be no Transocean were there not a BP seeking to extract ancient oil from the deep earth. There would be no Halliburton were the BPs of the world not in need of plugging their holes with cement. BP—and ExxonMobil, and Chevron, and Shell, and the countless other oil companies around the world—create and control the oil and gas industry. Everyone else is a hired hand, doing the dirty work.

There would be, over the months ahead, more attempts by the interested parties to brand the disaster this way or that. BP would refer to the Deepwater Horizon response, or the Gulf of Mexico response, while Transocean would refer to the "Macondo well" incident. The presidential commission investigated the "BP Deepwater Oil Spill," while the White House called it the "Deepwater BP Oil Spill."

BP was a brand-conscious corporation, having changed its name twice since the late 1990s, first from British Petroleum to BP Amoco, then to just the two letters. It had pushed the notion that it was a "green" oil company, one that accepted the scientific consensus on climate change, invested in alternative energy, and was looking "Beyond Petroleum." Along the way it changed its corporate logo from a shield to a Helios symbol, a green-and-yellow sunflower. But it couldn't shake a reputation for safety and environmental problems. In 2005, an explosion at a BP refinery in

Texas City, Texas, killed 15 people and injured 180. Four months later, a gulf hurricane severely damaged and nearly sank a huge BP production platform named Thunder Horse; in 2006, pipeline corrosion led to a major oil spill in Alaska. Hayward took over the company's leadership in 2007, vowing to focus "like a laser" on safety. Now this had happened. BP could blame other companies, but people were not confused on the issue of culpability: When Yahoo compiled the top ten search terms for 2010, right there at the top—ahead of World Cup, Miley Cyrus, Kim Kardashian, Lady Gaga, iPhone, Megan Fox, Justin Bieber, *American Idol,* and Britney Spears—was "BP oil spill."

A Giant Question Mark

There was a big problem with the early flow-rate estimate of a thousand barrels a day: A large oil slick had formed in the gulf, and it was growing at a rate that could not possibly be explained by a modest leak on the seafloor. You could see this slick from space. It swirled in the northern Gulf of Mexico like something tossed by the Mississippi's bird's foot. It was shaped a bit like a giant question mark.

The slick didn't look to John Amos like something coming from a manageable, tame, thousand-barrel leak. He's a geologist, and president of the nonprofit organization SkyTruth, based in Shepherdstown, West Virginia. The group uses publicly available satellite imagery to track oil spills and other environmental hazards. As soon as Amos heard about the Horizon explosion, on the morning of April 21, he started scrutinizing data and images from government satellites. He published maps of the spill on the SkyTruth blog.

On April 25 an image obtained by a NASA satellite showed a huge oil slick that dwarfed the skimmer boats at its periphery. By Amos's calculation, an oil sheen had already covered 800 square miles of the gulf. It was expanding at astonishing speed. In two days—from April 25 to April 27—the slick tripled in size.

Ian MacDonald also didn't buy the low flow-rate estimate. MacDonald, an oceanographer at Florida State University who has specialized in finding natural oil seeps and estimating their size using satellite imagery, saw the sprawling slick and knew that Macondo had to be more than a modest leak. On Monday, April 26, he called Amos and said the official numbers must be bogus.

"I agree, but how do we prove it?" Amos asked.

MacDonald had a postdoc in his laboratory, Oscar Garcia, who for his dissertation had developed a computer algorithm for calculating the amount of oil on the surface of the water via satellite imagery. It was a much faster technique than the one used by government scientists. On Wednesday, April 28, Amos directed MacDonald to a Coast Guard map of the oil that showed not only the extent of the slick but also described the appearance of the oil, the texture, and the character, in different locations. In some places, there was only a silver sheen; but in other regions, there was "rainbow sheen," and in other places, a "red-orange emulsion." Wielding the mouse of his computer, MacDonald outlined on his computer screen the various regions of oil, then assigned, based on the Coast Guard description, an estimate of the thickness of the oil in each region. It was a guess, no question. But this was also the subject of MacDonald's lifework—he was pretty much the singular expert in the field—and so it was a highly educated guess. He created a spreadsheet. Finally, totaling the different numbers, he came to a figure:

Nine million gallons.

That came out to 26,500 barrels a day.

What did I do wrong? MacDonald asked himself.

Had he not been conservative enough in his assumptions? Did the authorities—the government, BP, and so on—see something he didn't?

Journalists soon began calling him up. He revealed his flow-rate estimate.

"I was stuck with what I said," MacDonald told me later. "And then I kind of got worried. I thought about all the ways it could be wrong. I was desperate to see the video of the leak."

His number was actually a minimum, because it didn't take into account evaporation or other ways the oil could be naturally dispersed. If he was right, Macondo was gushing *at least* 26,500 barrels a day.

In that same stretch of late April, scientists in the Seattle office of the National Oceanic and Atmospheric Administration came up with a new estimate of their own, based on the appearance of the slick on the surface: 5,000 barrels a day.

BP was busy with its own set of calculations. The company crafted three scenarios based on the appearance of surface oil. The three estimates were labeled low, best guess, and high. The company categorized the oil as "sheen," "dull oil," and "dark oil," and estimated the extent to which each type covered the gulf surface. Each scenario assumed different thickness for the oil and took into account evaporation, chemical dispersion, and skimming. Conclusion: a low estimate of 1,063 barrels a day, a best guess of 5,758 barrels, and a high estimate of 14,266 barrels.

In a separate calculation, based not on what was observed on the gulf surface but rather on what was known about the oil reservoir, BP engineers estimated that in the worst-case scenario, the well could produce 60,000 barrels of oil a day. But the company didn't think it was looking at a worst-case scenario, because the blowout preventer was sitting on the well doing . . . *something.* It was helping, surely. There was all that riser pipe, too, still inhibiting the flow a little bit, what with the way it kinked and twisted and pretzeled.

A company executive would mention the 60,000-barrel number to members of Congress in a closed-door meeting on May 3, but that number was essentially stamped Do Not Believe, and BP gave preferential treatment to its "best guess" of 5,758 barrels. Which came pretty close to matching the NOAA estimate. So the Unified Command, on May 28, decided to go with the figure of 5,000, a nice round number.

This represented a 400 percent increase from the previous estimate. On its own, this increase in the estimated flow rate would have generated news stories. But something else was going on in the deep gulf that was even more alarming: The gusher was evolving.

Footage from an ROV, taken April 22, had showed nothing leaking in the vicinity of the blowout preventer. But on April 28, technicians flying one of the ROVs saw that a leak had formed, a new plume, right where the riser was kinked. Two leaks had become three. How many more would there be? Clearly, this wasn't a static event. This showed every sign of eroding from bad to worse. The blowout preventer could fall over, for gosh sakes, or there could be a second blowout at depth that would turn the wellhead into a crater.

That's what happened thirty-one years earlier in the southern Gulf of

Mexico, in the Bay of Campeche, when the Ixtoc I well blew out. The industry needed ten months to kill the well. That was the worst accidental oil spill ever, surpassed only by Saddam Hussein's mad destruction of Kuwaiti oil fields during the Persian Gulf War.

The dramatic corrosion of the riser changed the mood of the response. The administration wisely decided to get out in front of what was sure to be big breaking news. Administration officials in the White House Situation Room contacted the president on Air Force One and told him of the additional leak. Landry and Suttles convened an unusual late-evening news conference on the twenty-eighth. Both looked tense. Landry went first, starting with a bland comment about preparations to minimize the ecological impact of the spill, and then, finally, dropping the bomb:

"BP has just briefed me on a new location of an additional breach in the riser of the deep underwater well. While the—while BP believes, and we believed, and established a thousand-barrel-per-day estimate of what is leaking from the well, NOAA experts believe the outcome could be as much as five thousand barrels."

Landry then assured reporters, "That being said, BP has always anticipated and planned for a much larger spill."

Suttles took the microphone.

"Since the Transocean Deepwater Horizon incident began"—making sure to brand it right—"we've consistently provided frequent updates."

Suttles focused on the accomplishments:

"We successfully applied 42,000 gallons of dispersant. This almost doubles yesterday's total, which doubled the day before. We collected over 12,000 barrels of oily water, which represents a fourfold increase than the day before . . ."

Doubling, quadrupling—it sounded like progress. But it also sounded like Vietnam. The public can tell when the authorities are trumpeting the body counts even as they are losing the war.

Anyone could do the math. If it was 5,000 barrels a day, that was 210,000 gallons of oil every day pumping into the gulf. (There are 42 gallons in a barrel; the gallon-barrel distinction is one that proved confusing for many of us in the press, and I personally had to run two corrections on stories. The second time, when I wrote "4 million gallons" when I meant "barrels," a reader lectured me at length for, in effect, covering up 98 percent of the disaster. My only defense was that, the first time I made the mistake, and wrote "50 million barrels" instead of "gallons," I had overestimated the size of the spill by 4,100 percent.) The *Exxon Valdez* was typically described as an eleven-million-gallon spill. That meant that, if the well gushed for two months, the Deepwater Horizon spill would eclipse the *Exxon Valdez* spill. In any case, the story now fit into the *Exxon Valdez* oil-slick template.

And people knew what that looked like. Oiled birds. Oiled beaches. A horrible mess.

This was the oh-shit moment of the crisis. The news media recalibrated; more reporters jumped on planes and raced to New Orleans. Macondo had been a big story already, but now this was clearly a whole new level of crisis, and Deepwater Horizon suddenly became the biggest story in America.

This oil spill, President Obama would soon tell the nation, trying to convey his comprehension of the gravity of the situation, "could take many days to stop."

The Obama administration quickly recognized the other disaster to which this oil spill might soon be compared: Hurricane Katrina.

Never mind the similar location. Like Katrina, the oil spill would test the flexibility and creativity—and the basic competence—of the executive branch in the face of a sudden calamity. People would judge Obama and his aides on how they handled this disaster. The president knew it, the aides knew it, the news media knew it, and the president's political opponents knew it.

"Obama's Katrina" became a stock phrase in the news media and among critics of the administration, some of whom were still rankled by what they deemed unfair treatment of President Bush in the aftermath of the great storm of August 2005. One prominent Republican in Congress, Representative Darrell Issa of California, would spend the coming weeks blasting the administration for not doing enough fast enough. So too would the Republican governor of Louisiana, Bobby Jindal, sometimes named as a potential presidential candidate. The parish presidents in Louisiana, led by the perpetually dyspeptic Billy Nungesser, were furious about what they saw as a sluggish federal response.

What this political climate meant for the Obama administration was that it would not only have to handle the disaster effectively but also would need to be perceived as handling it effectively. Obama had to get all the right people in place, doing all the right things, and he also needed to appear to be fully engaged. Image management isn't a crime, it's sensible politics—and necessary to the extent that it would bolster public confidence in the response.

The lingering image of President Bush during Katrina was the photograph—perhaps the most ill conceived in the history of presidential image making—of the president peering out the window of Air Force One with a look of concern on his face as, somewhere below, unseen, were the victims of the hurricane. Obama would not repeat that mistake.

On Thursday, April 29, the administration threw together some high-profile briefings for reporters. Obama addressed the crisis in the Rose Garden, saying that he might call on the Pentagon to help. The secretary of homeland security, Janet Napolitano, announced in the press room, "Today I will be designating that this is a Spill of National Significance." This was a concept developed after the *Exxon Valdez* spill, and it was the first spill ever to merit the formal designation.

On Friday the thirtieth, Cabinet secretaries began parachuting into the Gulf Coast (not literally, but in retrospect, that might have been a good public relations maneuver). Ken Salazar told reporters, "We cannot rest and we will not rest until BP permanently seals the wellhead and until they clean up every drop of oil." He was gradually developing his sound bites and finally came up with a Wild West metaphor, offered up that Sunday, May 2, on CNN: "So our job is basically to keep the boot on the neck of British Petroleum."

President Obama changed his weekend plans and flew to New Orleans on Sunday afternoon. He traveled to Venice, the staging area for the Coast Guard, and made a statement while standing in a soaking downpour. An aide later insisted to me that this was not a stunt, and that the president of the United States truly did not have at his disposal an umbrella. In any event, the sodden president proved to be the perfect reciprocal to Bush-in-the-window.

Obama said the spill was a "massive and potentially unprecedented environmental disaster," and "that's why the federal government has launched and coordinated an all-hands-on-deck, relentless response to this crisis from Day One."

Relentless. From Day One. That became the administration mantra.

The administration rummaged the thesaurus for ways to convey the relentless, aggressive, vigorous, unflagging, indefatigable efforts to save the gulf.

Obama: ". . . aggressive search-and-rescue effort . . . we immediately and intensely investigated by remotely operated vehicles the entire 5,000 feet of pipe that's on the floor of the ocean. In that process, three leaks were identified . . . we've made preparations from Day One . . . immediately set up command center operations . . . ensure that we are doing whatever is required . . . from Day One, we have prepared and planned for the worst . . . while we have prepared and reacted aggressively, I'm not going to rest . . . or be satisfied until the leak is stopped at the source."

The news bulletins focused on what he said next: "Let me be clear: BP is responsible for this leak; BP will be paying the bill."

Then he went back to "our focus now is on a fully coordinated, relentless response," and so forth, and thanked the thousands of people who were working to stop the crisis, "whether it's the brave men and women of our military, or the local officials who call the gulf home."

The rainy-day message set a template for all of Obama's comments throughout the summer: Obama would be selective in his description of the people responding to the spill. The government scientists, members of the military, and ordinary citizens would be acknowledged, but the employees of BP, Transocean, Oceaneering, and the other oil industry companies participating in the response would be elided from Obama's statements. Obama wouldn't even mention them obliquely. He verbally airbrushed them out of the picture. Only someone who understood how the Unified Command was set up would know that it was not the government that used the ROVs to discover the three leaks. That was the work of BP and its contractors.

Yet everyone—certainly including those of us in the news media—was just now getting up to speed on the situation, struggling to master the language of the drilling industry, and trying to comprehend the relationship between public and private entities in the Unified Command. The whole situation was confounding, this notion that the government would be in charge but that BP would be the responsible party. No one could understand it. It was inherently awkward: Every government official wanted to kick the crap out of BP. The Unified Command had elements of a shotgun wedding. The federal government was stuck in a second-rate buddy movie in which the hero and the villain are chained together at the ankles.

The administration soon released a timeline documenting its relentlessness in responding to the blowout and the spill. The document was a festival of adverbs, each one laboring to invigorate the government response:

> The US government response to the BP Oil Spill began immediately after the explosion on the night of April 20 as an emergency search-and-rescue mission . . . Concurrently, the administration also quickly establishes a command center on the Gulf Coast . . . the administration has continuously anticipated and planned for a worst-case scenario . . . The President is alerted to the event and he begins actively monitoring the situation . . . Interagency coordination begins immediately among federal partners . . . The Regional Response Team immediately began developing plans . . . The Coast Guard continued to actively search for all 11 individuals still missing through the night.

The danger with producing a really detailed timeline is that the dispassionate observer might wonder if *the effort to detail the effort* may soon become greater than the effort itself. But the criticism of the administration was clearly unfounded: It had acted quickly, scrambled the jets, and scaled up the response appropriately with each new development in the gulf. No one could have imagined when the well exploded how bad this would get. This was unprecedented. This was, to use the fashionable term among scholars of crisis, a "Black Swan" event. Should people have known from the get-go that the rig would sink and the riser would gush massive amounts of oil and a slick would cover thousands of square miles of the gulf with red-orange mousse and rainbow sheen? Only in a parallel world where clairvoyance is the norm.

Had I been Obama's speechwriter, and did not care about such trivialities as whether I would be fired within minutes, or whether my boss would be reelected, or, worse, summarily impeached, I would have penned the following words for the president to utter as he stood in that Venice downpour:

> A little more than a month ago, I announced a new energy policy for this administration premised on the notion that offshore oil drilling has reached a level of technological maturity such that it could be presumed to be safe. But I wouldn't know a blowout preventer from a plumber's helper. I wouldn't know a hydraulic hot stab from a Wiener schnitzel. I wouldn't know a blind shear ram from a deaf-mute goat. As president, the buck stops with me, but let's face it, my staff let me down. I pay people to figure out the technical stuff. I'm a lawyer! I don't know how they poke holes in the earth and extract crude oil, and I'm not so sure I should have to know. Arguably, that's way, way below my pay grade.
>
> So now I'm all wet—literally, unfortunately, as I stand here amid staffers who can't even find an *effing umbrella*—and I have to tell you, the American people, the painful truth. We're kind of groping our way forward here. We have a hole in the bottom of the Gulf of Mexico, and we, the government, don't have the tools to fix it. We're counting on BP to solve this problem that they created; if they don't fix it, we'll yell and shout, but mostly we'll be shit out of luck, pardon my French. For years the only weapon the federal government has wielded against the oil industry has been a rubber stamp. Now the oil industry has done to offshore drilling what the financial industry did to the economy. It's been a free-for-all, and the inmates have burned down the asylum. I wish it were otherwise and wish I could tell you that the government can solve this problem. I also wish we did not need, as a nation, to punch holes in the gulf. We have an addiction to oil. Yeah, feel outraged by what BP has done, but if you really don't like deepwater drilling, probably you shouldn't drive. I'll get back to you when I have more information. In the meantime, God help us.

Chapter 5

Doomsday

The president needed to demonstrate that he was doing more than just lobbing tough words at BP and throwing a few Cabinet secretaries at the problem. The government had to make a difference. The proper role of government may be a source of partisan rancor and division when discussed in the abstract, but in a crisis, most Americans expect the US cavalry to appear at the ridgetop. Even the Republican governor of Louisiana, Bobby Jindal, an ideological critic of Big Government, protested that the Obama administration wasn't doing enough to combat the oil spill.

Obama's government might not have any Macondo-plugging hardware, but the president had at his disposal a vast federal apparatus that included some of the smartest people in the country.

Clearly, it was time to call in the scientists.

The national laboratories run by the Department of Energy boasted some of the best scientists and engineers in the nation. Within two days of the discovery of the third leak on the gulf floor, top officials at DOE sent an email to the labs: Help us. We need ideas. Come up with a plan for killing this oil well. Be inventive. Think.

Three national laboratories—Sandia and Los Alamos, in New Mexico, and Lawrence Livermore, in California—are tasked with a number of grave national security tasks, most notably the designing and testing of the country's nuclear weapons arsenal. Los Alamos was founded as a secret research facility for the Manhattan Project during World War II. The Department of Energy technically owns all of the country's nuclear warheads; the US military borrows them. (Presumably these are not casual transactions—one doubts that anyone from DOE calls up one day and says, "We want our nukes back.") Through their work on explosives, the DOE scientists and engineers understand the physics of kinetic events, how shock waves propagate, how steel withstands stress. If they could understand the awesome power of the "strong force" that binds the constituents of an atom and gives nuclear weapons their obliterating energy, they surely could figure out what to do with a runaway oil well in the Gulf of Mexico.

Someone who did not understand the teeth-grinding anxieties of the nation during the oil spill might have perceived this appeal to the national labs as a sign of panic. It was, to be sure, a little bit like having car trouble one morning in the driveway and making an emergency call to the people at NASCAR. If your computer blinks out with the Blue Screen of Death, you don't call directory assistance in Seattle and ask for Bill Gates's phone number. But Macondo wasn't like other problems. Obama needed to get his biggest brains in the game.

At the Sandia National Laboratories in Albuquerque, Tom Hunter suddenly found himself on oil spill duty. Hunter ran the lab. He was not wholly unfamiliar with the oil patch; during college he'd been an oil field roustabout in Louisiana. Later, at the University of Wisconsin, he received his PhD in nuclear engineering. He'd had a full and productive career, and was days away from retirement. When he first heard about the rig fire, he told me later, "I didn't think of it as an oil spill. I did not have this concept that the oil was coming out." He figured that out soon enough.

For a couple of days, the Sandia scientists knocked around ideas for killing the well and kept coming back, Hunter said, to what they knew best: explosions. They could blow up Macondo. Not with nuclear weapons, of course, but with some shaped charges—purely conventional. They could detonate something at the wellhead or down in the depths of the well to crush the pipe and/or collapse the hole. They could go kinetic, hit it hard, and end this nightmare with a bang.

The idea of using explosives to solve the oil leak would become, in the days and weeks ahead, a kind of twitch that no one in the scientific or engineering community could completely suppress. Word spread that the Soviet Union dropped nuclear weapons on blown-out wells back in the day; further investigation suggested that the explosives had been used only on land and never on offshore blowouts. The bomb-the-well idea had a hitch, however: Everyone involved in the response, both within the government and at BP, wanted to make sure that no action would exacerbate the spill. "First, do no harm" was the governing philosophy. Don't try to fix a hangnail with a chain saw. Don't try to bathe the dog with a fire hose.

Adding all that kinetic energy to Macondo to seal the skinny hole in the earth might wind up opening countless other fissures that would let the reservoir flow into the gulf waters. Macondo could end up dumping its entire load. The existence of oil reservoirs is made possible, after all, by seal rocks, and although the seals are not always immaculate (natural seeps of methane and oil are common), in the case of Macondo, the rock formations had kept a large quantity of oil and gas buried in the earth since before the first protohuman came down from the trees.

Hunter's first conference call with scientists and government officials, held on Friday night, April 30, ended inconclusively. They were all thrashing around. Hunter convened another conference call the following morning, but they still didn't have a workable idea.

"We concluded that this was probably not an explosives problem at all," Hunter said. "What we needed to do is find out more about it."

The smartest people on the federal payroll knew next to nothing about deep-sea oil wells, petroleum engineering, blowouts, blowout preventers, ROVs, hot stabs, and so on. They knew how to turn heavy elements into explosive devices, but had no idea how to turn off an undersea faucet.

This would be a learning experience for everyone, with the oil spill as a powerful accelerant.

So now Obama had his best minds working on the spill. What the president really needed was someone to be the public face of the government's response. He needed someone to play the role of hero to the rescue. But this wasn't that kind of disaster, and it probably never would be. From the very beginning, the crisis had a conspicuous deficit of heroes, as if some chemical in the spill repelled anyone who might solve the problem. This wasn't like Apollo 13, where three astronauts in a damaged spacecraft shivered their way around the moon and somehow, through courage and guile, and aided by the ingenuity of engineers back in Houston, found their way home. There were battalions of heroes that time. But this time, failure *did* look like an option. Failure was, at the very least, the default position pending further developments.

Politicians and oil executives discovered that the oil spill slimed everyone who came into contact with it. It didn't matter if you performed splendidly; as long as the well remained ungoverned and ungovernable, the American people would be angry, disgusted, and anxious. The government could cite statistics all day long about how many feet of plastic boom had been laid to stop the oil from coming ashore, and how many boats were skimming oil, and how many planes were flying how many sorties to drop how many gallons of chemical dispersants on the slick. What people wanted to hear was that it was over. They didn't want amelioration, they wanted termination.

Obama did have one ace in the hole. He had on his executive branch payroll a person who seemed to be made for this kind of moment: Thad Allen, the Coast Guard commandant, and, as of May 1, the National Incident Commander for the BP oil spill.

Allen radiates unflappability. Some of this is his appearance, and the natural instinct of observers to assign positive traits to people of a certain stature and demeanor. Allen stands straight and wide, like a wall, a human bulwark. He is just framed that way, with heavy lumber, as if whoever put him together knew that he would someday have to weather serious storms.

Even by Obama standards, Allen possesses an unusual equanimity. He does not speak in exclamations. From the start of the crisis, he set a certain tone of all-business. He would not craft sound bites for the media mob. He had a goal, which was to get to the end of it. Keep the end in mind. Focus on the objective. This was Clausewitz on the Gulf of Mexico. He wanted no part of pyrrhic political victories. When a BBC reporter asked him if the government had considered firing BP from the oil spill response, Allen stiffened momentarily and said, "I *am* in government, and there was no solution without BP."

He has a technique, he told me, of visualizing what he will say and how he will say it before he gets in front of the microphones and cameras.

It's similar, he said, to how an athlete visualizes a play. "Once I get the mental model down, and understand the mental landscape, my job is to craft a way to the end. That was my job, every day. Create the art of the possible," he said.

Somewhere along the way, he became that rarity in modern American politics: the nonpartisan figure, not obviously ideological in any way. He wasn't, after all, a political appointee. Allen was a sailor who worked his way up to the helm of the Coast Guard. He earned his stature as the nation's go-to guy for crisis management. He cultivated the skill of remaining calm even when other folks had the bulging eyeballs, the throbbing forehead veins, and the sweaty palms. He had charged to the rescue when the federal government bungled the initial response to Hurricane Katrina. He'd handled oil spills and had even gone through a training exercise, back in 2002, for handling a major Gulf of Mexico oil spill. As commandant of the Coast Guard, Allen had started working on the Deepwater Horizon disaster within hours of the incident, staying up through the night to monitor the search and rescue operation. On Day 12 of the crisis, after it was clear to everyone that this was a different level of bad, his boss, Janet Napolitano, appointed him National Incident Commander, which gave him his own personal fiefdom—the National Incident Command—separate from the bureaucratic framework of the Unified Command. (This kind of flowchart stuff may seem esoteric, but Allen could recite it off the cuff, and when I visited him in his small, windowless National Incident Command office at Coast Guard headquarters, the only two things on the wall were flowcharts of the command structure.)

This was different from Katrina. "One of the differences between a hurricane and an oil spill is you have someone to blame," he said. The party to blame was obviously BP, but that didn't mean Allen was going to shun the company or push it away. He had to keep the federal government and BP moving in the same direction, toward a resolution to the crisis. The government and BP would never be a team, exactly, but they had a common enemy in the oil spill. This alliance, or whatever you wanted to call it, would become trickier as the weeks went on, the spill remained unabated, and the political rhetoric sparked and sizzled. The admiral needed to have open lines of communication with the top executives at

BP even as, back in Washington, Ken Salazar would say something like "We have them by the neck"—a statement the interior secretary delivered as he used his hands to choke an imaginary oil company.

Allen drew the analogy to being in a car crash that injures your child. The other driver caused the crash. But the other driver turns out to be a doctor.

"Are you going to say, 'Don't treat my child?'"

Allen is a fast talker, and comprehending his words can be hampered by his tendency to sling jargon at a hundred miles an hour. He also favors the fifty-dollar word in special circumstances, such as: "This spill at this point in my view is indeterminate. That makes it asymmetrical, anomalous, and one of the most complex things we've ever dealt with."

He wasn't campaigning for anything, didn't have political aspirations, so he could tell people the truth with little or no varnish. He could be honest about what the government could and couldn't do. The military, for example, couldn't fix this. Only the oil industry could solve this problem. "The private sector owns nearly all the means to deal with this problem and fix the leak and stop the source," Allen said in an interview early on.

The admiral had thirty-nine years of public service behind him, but he was scheduled to relinquish command of the Coast Guard on May 25. The rules required that he retire from the Coast Guard if he wanted to keep his four stars. From May 25 to June 30, he was supposed to be on leave.

He and his wife, Pam, were going to buy a piece of property and build a new house. They would take a two-week vacation to Ireland in early August.

That, at least, was the plan.

At the seafloor, amid the blizzard of marine snow, the suddenly world-famous ROVs scrutinized the blowout preventer. Which was also world famous now—or world infamous. The BP executives wasted no opportunity to assign the blame for the April 20 tragedy to this Transocean-owned, Transocean-maintained stack of valves, but they clung to the notion that the blowout preventer might yet function. Maybe it had a magic bullet hiding in its hardware.

On Monday morning, May 3, good news burst from the gulf: "We've significantly cut the flow through the pipe," BP spokesman Jeff Childs said at a briefing of Alabama officials in Mobile. According to Childs, the ROV intervention had succeeded in partially closing one of the annular preventers, a big diaphragm-like device near the top of the stack. The country, fatigued after so long a crisis—this was Day 14—had to be thrilled at such a positive development. Unfortunately, it wasn't true. The BP spokesman correctly reported that an ROV had managed to activate the annular preventer, but the valve didn't seem to have the heart to keep the flow clamped.

People would come to learn a guiding principle about Deepwater Horizon oil spill information: If it contained good news, it wasn't true.

That afternoon, Rear Admiral Mary Landry and BP's Doug Suttles held another Unified Command news conference in Robert, Louisiana, and Landry once again spun through the many good things happening with the response:

"We continue to aggressively execute and adapt our local and regional response plans to the on-scene conditions in accordance with the preexisting, established, and exercised plans as required by the Oil Pollution Act of 1990," Landry said, calming the untold millions of Americans worried that the government was not properly following the procedures outlined in the Oil Pollution Act of 1990. She reported that BP had developed an "innovative technique" for applying chemical dispersants directly to the well, and she noted that the president and Admiral Allen had talked to local officials, "promising them continued and unparalleled coordination and to spare no expense in this intense and unrelenting response effort."

After several minutes of this rah-rah talk, and another minute or so of an MMS official speaking in bureaucratese, Suttles took the microphone. Straight off the bat, he confirmed the bad news that the morning story from Mobile was wrong, and the BOP intervention hadn't curtailed the leak.

"You would see me doing cartwheels down the hallway if that were actually the case," Suttles said.

Suttles was not the type to emote. He didn't stray beyond the immediate engineering issues of the day. But sometimes, as on this morning, he let a human emotion slip into the mix—or, to be more precise, he acknowledged that, hypothetically, a human emotion might be encountered during the problem-solving process. There could be engineering achievements of sufficient magnitude to incite a celebration, in theory. Although it must be said that Suttles did not seem like the kind of person who had done a cartwheel in quite a while.

The magic of modern technology enabled the engineers in BP's war room in suburban Houston to watch, in real time, what the ROVs were doing on the floor of the gulf. But the engineers came to realize that there was something fishy about the subsea hardware. That blowout preventer they were looking at through the video feeds did not precisely match the diagrams of the blowout preventer that they'd been using in the war room. For example, an ROV inserted a hydraulic line into a port that was supposed to be connected to the middle pipe ram. The middle pipe ram could, in theory, seal the well. But the engineers came to understand that they were poking the wrong hole. That port was not, in fact, connected to the middle pipe ram. It was connected to a "test" ram at the bottom of the stack. The test ram was useless: It was good for certain kinds of diagnostic operations but couldn't seal fluids coming from below.

"This stack is plumbed wrong," Billy Stringfellow, one of the Transocean engineers, said in the war room.

BP engineer Harry Thierens wrote in his journal, "When I heard this news, I lost all faith in this BOP stack plumbing."

This BOP was not only misplumbed, it had leaks and weak batteries. It was a jury-rigged device. It had been tweaked, poked, prodded, rearranged, and reconfigured more often than an aging movie star. The records of the changes by Transocean were hard to retrieve, or had gone down with the rig, or perhaps had never been recorded at all. In the deep sea, the blowout preventer had become an enigma, a five-story stack of anomalies.

The engineers despaired of this blowout preventer. On May 5—Day 16 of the crisis—BP officially brought to an end the "BOP intervention" portion of the disaster response. The calamity accounting now stood at one blowout with fatal fire, a sunk rig, two subsea oil leaks, a third subsea oil leak and a much higher flow rate estimate, and a failed BOP intervention, with each of these things happening sequentially to torque the crisis to an ever more painful position.

The oil slick, now the size of a small state, continued to grow, creeping toward shore, toward the delicate barrier islands of the Louisiana coast and the fragile marshes of the Mississippi River's bird's foot delta.

Government officials would now have to fight the oil with water: by opening reservoirs upriver, and opening the diversions from the river channel, letting all that fresh water push into the brackish marshes and, in effect, push the oil spill away as much as possible. But this would work only to slow down the advance of the slick. It was going to come ashore eventually. In the water versus oil fight, oil would find a way to win.

BP still had plenty of options. It could try to capture as much oil as possible as it came out of the well. The company prepared its next gambit: the deployment of something called the cofferdam that could be lowered onto the big leak at the end of the collapsed riser.

The relief wells offered the only long-term fix. The first relief well had been spudded on May 2. The drillers planned to spiral in toward the base of Macondo, intercept the well, pump mud down the relief well, send in a thick slug of cement, and end the spill once and for all. Macondo would be plugged at the bottom, and, after some final abandonment and capping procedures up at the wellhead, everyone could go home. But the relief well wasn't a gimme. The target—a spot on the final, skinny string of steel casing near the bottom of the well—was just seven inches across, not to mention more than three miles below the gulf surface.

The engineers figured they would succeed by August.

Doomsday

Just as the American citizenry was trying to get a grip on how bad this oil spill would be, a slew of experts rushed forward to assure the country that the situation was definitely far worse than most people could possibly imagine.

The scaling-up of the crisis and the intensification of the horror into an event of extreme national anguish would require an expansion of the geography of the disaster. As long as the spill and its consequences were confined to the unloved, skinned-knee Third Coast, which had already been the dumping ground of the United States, the disaster would have a regional feel to it. It would be, for most Americans, something that happened elsewhere. This wasn't affecting a place where the political and media elites—the curators of national angst—owned their vacation homes. It's not like this was the Hamptons.

But then the experts emerged with a new and terrifying concept: the Loop Current.

"Oil Set to Hit 'Loop Current': Would Send Spill to Florida Keys."

So read the screaming banner headline on the home page of the Huffington Post. The Loop Current had existed for countless eons but had somehow eluded the attention of the general public until the BP oil spill. The oceanographers said the Loop Current could transport the oil from the Gulf of Mexico to the Straits of Florida and then, amazingly, hand it off—as one sprinter passes a baton to another—to something called the Florida Current, which in turn would pass it to the Gulf Stream. The Gulf Stream would carry the oil up the Florida coast so close to the shore that it could be seen by Palm Beach aristocrats shopping on Worth Avenue. The oil would ride the Gulf Stream, bump into the Outer Banks of North Carolina, and carom toward Bermuda. The TV networks carried a report, based on a NOAA study of currents, showing the oil from the gulf circulating in a massive gyre in the central Atlantic—a cyclonic superstorm of Macondo crude.

Conceivably an element of alarmism had wormed its way into this emerging narrative. But credible academic oceanographers seemed genuinely concerned about the potential for ocean currents to carry the Macondo oil around the Florida peninsula. An oceanographer at the University of South Florida said that if the oil got caught up in the Loop Current, it would take about a week to reach the Florida Straits, then another week or so to arrive off the beaches of Miami and Palm Beach, and finally another week, roughly, to get close to Cape Hatteras, North Carolina. A University of Miami oceanographer said it was "inevitable at this stage" that the oil would get into the Loop Current, and he added that if the well continued to spew for two or three months, "it'll be catastrophic."

I duly reported this in a story that ran on the front page of *The Washington Post:*

Scientists envision devastation for gulf

Much is still unknown, but 'Loop Current' could take spilled oil all the way to N.C.

The scientists emphasized their uncertainty. Every scientist understands that results are vulnerable to subsequent correction or falsification. Truth is rarely absolute. In the case of the oil spill, the oceanographers could only say with certainty that the oil, *if* caught up in the Loop Current, would *likely* take the long ride around Florida.

And yet the scale of the disaster, the hunger for information, created a powerful market for hurried-up science, for science-on-the-go. In the weeks to come there would be boat-deck science—research results announced by satellite phone from the open sea within hours of instruments coming out of the water. There's no time for peer review when the world is coming unglued.

Ian MacDonald, the Florida State oceanographer who had realized the spill was larger than officially estimated, felt uncomfortable with this accelerated process and worried about his own role—the way he'd given back-of-the-envelope numbers to the news media without the tempering effects of peer review. But there didn't seem to be time to waste. This was not an ordinary situation.

My May 5 *Post* story tried to call attention to the uncertainties:

> The crisis in the gulf is shot through with guesses, rough estimates, and murky figures. Whether the oil blows onshore depends on fickle winds. This oil slick has been elusive and enigmatic, lurking off the coast of Louisiana for many days as if choosing its moment of attack. It has changed sizes: In rough, churning seas, the visible slick at the surface has shrunk in recent days.
>
> The oil by its nature is hard to peg. It's not a single, coherent blob but rather an irregular, amoeba-shaped expanse that in some places forms a thin sheen on the water and in other locations is braided and stretched into tendrils of thick, orange-brown gunk. There may be a large plume of oil in the water column, unseen.

All fine, except the two uncertainty paragraphs were tucked far into the story, past the Loop Current shenanigans. And, given that the whole thing was packaged under the headline "Scientists envision devastation for gulf," the uncertainty paragraphs served as a weak expectoration of equivocation amid the maelstrom of certain doom.

Human nature does not give itself easily to the calm cataloguing and analysis of risk. What people want to know in a crisis, first and foremost, is whether they should run for their lives. Rational people now felt stalked by the oil. They no longer saw the oil as a passive agent. It had evolved. It didn't just spill stupidly into the water and float there, but rather had become an active force of mayhem, as mobile as a serial killer. The spill prowled the watery margins of the nation. It would soon creep up the coast, infiltrate the bays, ooze into the rivers, work its way up the sewer lines, and then finally in the dead of night sneak into our bedrooms.

This was crazy talk, but the crisis had toggled into a new stage in which crazy was the new normal; crazy was the *baseline* of any reasonable discourse. If you weren't talking crazy talk, you weren't paying attention. The people who needed their medication adjusted immediately were the ones who thought this would probably turn out okay, that it was just 5,000 barrels a day, that the gulf was a big body of water, and dilution was the solution. Optimism, confidence, assuredness—these had become compelling signs of derangement.

The crisis provided fertile ground for the germination of bad information and outright hoaxes. "On the night of April 20, the North Korean Mini Submarine manned by these '*suicidal*' 17th Sniper Corps soldiers attacked the Deepwater Horizon with what are believed to be 2 incendiary torpedoes," reported one Internet story. Most people could tell that was bogus, but it was harder to interpret the report that the Russians had visited the well in their special manned submersibles, or the report that a huge quantity of methane gas lurked just beneath the gulf floor and could at any moment bubble upward, expand violently, and create a gaseous hole in the sea into which ships would just . . . fall. The ships would plummet into nothingness, swallowed whole. The sea would slurp them down like oysters.

This disaster was starting to feel biblical. Before one of Governor Jindal's news conferences, a New Orleans pastor, Randy Craighead, prayed for divine intervention: "Father, we pray for a prevailing north wind to drive that oil slick southward."

* * *

No one rang the doomsday bell more passionately than Matt Simmons, a prominent and provocative figure in the oil industry. The founder of Simmons & Company, a major oil investment house, he had written widely on what he saw as the inevitable downturn in the industry as the petroleum reservoirs dried up and even the most intrepid oil companies ran out of places to drill. His 2005 book *Twilight in the Desert: The Coming Saudi Oil Shock and the World Economy* advanced the theory of "peak oil," which posits that the downturn in worldwide oil production, when it comes, will trigger global economic chaos. Somewhere along the way, Simmons went from pessimist to doomsayer. Appearing on Fox News Channel, he said the spill might be unstoppable. In a phone interview one day early on, Simmons told me, "They're going to have to clean up the Gulf of Mexico." He meant the whole thing: the entire gulf. And he saw BP as a hopeless cause.

"BP's history," he said. "I don't think they're going to be able to put the leak out until the reservoir depletes. It's just too technically challenging."

And Simmons didn't buy the estimate of fifty million to one hundred million barrels in the reservoir. He said there were a billion barrels down there. A billion barrels of oil were going to disgorge into the gulf and paint the whole thing black. When I asked him about the idea that the Loop Current would take the oil to North Carolina, he scoffed.

"North Carolina? *Ireland*."

The Cofferdam

Alarmism flourished in the absence of solid information about the scale of the disaster, and BP did not exert any energy to measure the flow of the well.

Early in the crisis, the company spoke with Rich Camilli and Andy Bowen, scientists at the Woods Hole Oceanographic Institution on Cape Cod, Massachusetts, which operates manned and unmanned submersibles. Woods Hole has instruments, Camilli and Bowen told BP, that measure the flow rates of deep-sea hydrothermal vents—those "black smokers" that are home to their own exotic ecosystems of blind crabs and bizarre tube worms. The Woods Hole sensors use acoustic signals to create a three-dimensional image of a plume. That technology presumably could be adapted to the Macondo well, offering a very precise measurement of the flow. Camilli and Bowen prepared to travel to the Gulf Coast to supervise such an operation. But then BP told them to stay put—the company's plan for a containment device to capture the oil was ahead of schedule and ready to be deployed.

When I asked a BP spokesman, David Nicholas, why the company hadn't attempted to measure the flow, he said it didn't matter, because BP was already doing everything it could to stop the leak.

"I don't think an estimate of the flow rate would change either the direction or the scale of our response to it," Nicholas said—even as, in Houston, engineers labored on a wide variety of schemes for which success or failure could depend on how fast oil was spurting from the well.

To critics of BP, there was an obvious explanation for what was going on: BP didn't want to know the flow rate because a really big number would be extremely expensive when the company eventually got socked with fines for polluting the gulf. The federal pollution laws under the Clean Water Act made BP vulnerable to a fine of $1,100 for every barrel of oil spilled. The fine could escalate to $4,300 per barrel if a federal court found BP guilty of gross negligence. This incident could potentially generate fines running into the billions of dollars. As long as the flow remained unknown, BP's lawyers would be able to sow uncertainty about the dimensions of the spill.

MacDonald, the oceanographer, echoed the thoughts of many of BP's critics: "BP was in the process of trying to hide the body."

The ROVs at the seafloor finally had a triumph, if a minor one. An ROV wielding a circular saw cut off the damaged end of the drill pipe jutting from inside the collapsed twenty-one-inch riser. The pipe had been spewing oil, the smallest of the three leaks from the well. After putting the clean cut on the pipe, the ROVs sealed it with a tight-fitting cap. The three leaks had become two leaks. The achievement did not reduce the overall flow of Macondo, because the hydrocarbons would simply come out faster from the other two leaks. But amid all the bad news, this gave BP a chance to say that something had gone right on the seafloor. The technology had gained some traction.

Next up: the cofferdam.

A cofferdam, also known as a containment dome, is essentially a very large box that can be fitted over an oil leak. A hole in the top allows the oil to be collected by a riser pipe and brought to a surface ship. There were several such cofferdams already in the Port Fourchon facility of Wild Well Control, the blowout-killing company run by the legendary Pat Campbell, who had helped put out the well fires during the Persian Gulf War.

BP picked out a four-story-high dome. The workers at Wild Well Control reinforced it with steel girders to keep it from being crushed by the deep-sea pressures. They added mud flaps, which jutted from the midsection and would keep the whole thing from sinking into the muck of the gulf floor. The cofferdam looked like something you'd use to stopper a giant bathtub. To the jaundiced eye, the containment dome did not scream Twenty-first Century Technology.

Ken Salazar, just three days removed from his Sunday talk show comment about keeping the boot on BP's neck, visited Port Fourchon and toured the compound at Wild Well Control. Accompanying him was a geophysicist, Marcia McNutt, the chief of the US Geological Survey. They walked inside the cofferdam as it sat on the dock. Salazar would make sure to mention this to reporters later—he'd gone *inside the dome*. He had been in the belly of the beast. Now, however, another glitch in the plan materialized. The workers needed to lift the dome off the dock and place it very precisely on the center line of an adjacent barge. If the dome was to one side or the other of the barge, there could be a load stability problem; no one wanted to see the cofferdam unceremoniously tumbling into the gulf halfway to the disaster site. The crane needed to be fully extended to place the dome on the barge, but the dome, after being reinforced with steel, weighed ninety-eight tons, too much for the crane at hand. Thus, the workers had to scramble to find a bigger crane.

McNutt's verdict: "That is a total rookie mistake."

The incident, she told me later, gave her a glimpse of what she came to see as the BP approach to almost everything. BP, she said, didn't like to spend a lot of time fretting about what might go wrong. McNutt described the BP attitude as "Let's do it and see if it works." This opened up an obvious role for the federal government as it tried to apply some oversight to the subsea response. The government could ask the hard questions, push BP to consider what might go awry, give some attention to the worst-case scenarios, and make sure that BP's plans weren't too driven by a git-'er-done attitude: "The government approach was, think carefully every step, what can go wrong, what's your worst nightmare, let's check everything thoroughly, let's model everything, let's make sure we know exactly what's going to happen before it happens."

Except this was a national disaster, and everything was moving at a heart-palpitating pace. The American people were screaming, Stop the spill *now!* So there was a natural tension between git-'er-done and let's-model-this-carefully.

The cargo ship *Joe Griffin* hauled the dome out of Port Fourchon, into the gulf, and putt-putted along for twelve hours before finally reaching the disaster site, known to the response team as the "source." The source was crowded with BP-contracted ships and boats and drilling rigs. An all-purpose servicing rig, the Helix *Q4000,* took the dome and prepared to lower it into the water.

BP executives informed federal officials that the probability of success of the containment dome was "medium high." At the BP headquarters, however, the engineers were anxious. They had chemistry on their minds. They worried about methane hydrates.

In the coming days, a nation that never liked chemistry class suddenly found itself hearing about mysterious, ice-like, slushy, crystalline substances that seemed to form magically in the deep ocean when hot, pressurized gas suddenly came into contact with cold, pressurized water.

An oil reservoir contains more than just oil. There's gas, water, sand. The stuff spurting out the end of the riser pipe was a concoction of coumpounds. The oil was itself a diverse array of chemical species. There is no single hydrocarbon molecule named "oil." There is, however, benzene, toluene, m-Xylene, n-heptylbenzene, indene, indan, naphthalene, tetralin, biphenyl, acenaphthylene, fluorene, pyrene, chrysene, benzopyrene, pentacene—these being just a partial list of typical aromatic hydrocarbons found in crude oil. There are also hydrocarbon gases: predominantly methane, but also ethane, propane, butane, pentane, hexane, and heptane. And there are other gases mixed in with the hydrocarbons—gases that have more than just hydrogen and carbon atoms in their chemical makeup. These include nitrogen, carbon dioxide, hydrogen sulfide, and helium. One also finds traces of phosphorus, iron, nickel, and vanadium.

Complex as this stew may be, a particular batch of crude can be described as light or heavy, and—depending upon the sulfur content—sweet or sour.

The Macondo crude was light and sweet.

Engineers at the refinery know how to tweeze apart the constituents of the crude and make the substances that fuel (or pave, or lubricate) modern society: gasoline, motor oil, jet fuel, diesel, paraffin wax, tar, roofing material, asphalt, and so on. But this process of transferring the oil and gas from the deep rock of the planet to the refinery is not supposed to include a detour through seawater.

The engineers were trying, with their cofferdam, to swallow up the hot oil immediately after it emerged from the riser. The danger was that the gas would immediately crystallize into hydrates and clog the dome. The engineers in Houston studied the numbers. This was a fifty-fifty deal, they reckoned. It was worth a shot. They could also ameliorate the hydrate formation by pumping hot water down the riser, the pipe that would bring the hydrocarbons to the surface. First, though, they'd place the dome on top of the leak.

On the evening of Thursday, May 6, Day 17 of the crisis, the *Q4000* lowered the dome through its moon pool.

The cofferdam eased beneath the waves.

The oil and gas billowed up.

Cofferdam, down.

Oil and gas, up.

They met several thousand feet above the seafloor. The methane hydrates began to form inside the dome. The dome hadn't even reached its target yet; it was still half a mile from the seafloor.

By the time the giant contraption reached the bottom on Friday evening, it was quite thoroughly clogged with hydrates. As the technicians, watching their operation via ROV cameras, attempted to place the dome over the big leak, the dome stopped dropping altogether. It had become buoyant. Then it began to rise. The dome was now moving back up the water column, toward the ships on the surface. Crammed with methane hydrates, the cofferdam had become, suddenly, not so much a solution to the spill as a threat to the surface ships.

The technicians got it under control and guided it to the seafloor some distance from the riser. It was a ninety-eight-ton failure.

"The issue was actually an installation process issue," BP engineer Richard Lynch told me later. Not everyone had understood the plan as devised in Houston. The dome was supposed to come in sideways over the plume, not get lowered straight down over it.

The cofferdam failure threw the BP engineers for a loop. They had talked about the hydrates but didn't realize how pernicious they would be. Lynch had successfully deployed cofferdams after Hurricane Katrina created oil leaks in shallow water. The deep water, clearly, was a whole different world.

BP would have to do better next time, take no chances, get everything storyboarded precisely and all parties fully up to speed. They needed to presume that hydrates would try to spoil every maneuver. They needed to elevate their game if they wanted to succeed in the deep-sea league.

Chapter 6

Calling All Geniuses

Tom Hunter at Sandia, and the directors of the Los Alamos and Lawrence Livermore national labs, each dispatched a scientist to BP's war room in Houston on Saturday, May 1. Hunter followed a couple of days later. On the evening of May 6, US Geological Survey director Marcia McNutt arrived, accompanying Salazar. For the secretary of the interior, this would be a mere drive-by visit of the BP war room operation, but McNutt would not be so lucky. She happened to know a lot about taking hardware into the deep sea, having served as director of the Monterey Bay Aquarium Research Institute (MBARI), widely viewed as the NASA of deep-sea research. Anyone knowing about deep water was suddenly needed on the front lines of the battle. Salazar took off and left McNutt in Houston, and she set up shop in a windowless six-by-ten-foot office. She had packed only a carry-on bag for her Gulf Coast trip.

The government scientists entered the spill response with an inherently awkward status. Their role was ad hoc. There was nothing in the National Contingency Plan about government scientists moving into an oil company's headquarters.

Then, on May 12, the alpha scientist showed up.

Steven Chu served Obama as the secretary of energy. He was a physicist and, as would be noted countless times in the days ahead, a winner of the Nobel Prize. For an administration that wanted to get its best people in the game, Steve Chu was a titanic figure, possibly the smartest man in the federal government, if not the known universe.

All previous energy secretaries had been politicians or businessmen; Chu broke the mold. When he worked at Bell Labs, he had figured out how to freeze atomic particles with lasers. That's what earned him the Nobel. He later joined academia, teaching at Stanford and Berkeley, and running the Lawrence Berkeley National Laboratory. Chu was a man who could tell you, off the top of his head, that a lithium ion battery stores 0.54 megajoules of energy per kilogram, while body fat stores 38 megajoules, and kerosene contains 43 megajoules (after which you would be permitted to say, off the top of your head, "What's a megajoule?"). In recent years, he had become particularly animated about the need to combat climate change through a dramatic transformation of the energy sector, and as Obama's secretary of energy, he had been doling out billions of dollars in stimulus-package money to boost the fortunes of alternative energy technologies.

He embraced solar, tolerated nuclear, hated coal. ("Coal is my worst nightmare.") He had warned in a Harvard commencement address the previous year that if the planet undergoes the kind of climate change that many models predict, the "change will be so rapid that many species, including humans, will have a hard time adapting." Anyone at BP doing a quick Google search on Steve Chu would quickly come to understand that he wasn't a bosom buddy of the petroleum industry.

Obama had asked Chu on May 10 to dive into the response. But Chu needed backup. He wanted to walk into BP's headquarters along with a team of first-rate scientists with unquestioned mental bandwidth. He checked in with the JASON group, a team of free-thinking scientists on retainer for the Pentagon to think about exotic threats to the nation. Eventually, calling around, he got his wingmen.

"I specifically didn't want oil experts. BP had the oil experts," Chu told me. He said he was looking for "someone who can become an expert very quickly, someone who is very, very smart. People who are willing to explore in very creative ways and not simply listen to, quote, 'the industry experts.'"

Chu and his staff told the scientific advisers to drop everything and fly to Houston that very night, May 11. They had a meeting with BP engineers scheduled for six thirty the next morning in Building 4 on the Westlake Park campus.

The Chu advisory team would henceforth operate in its own, parallel dimension—very much part of the response but not to be confused with the government scientists from the national labs. Of course the news media did get confused on this point. We knew that there was a "science team," led by Steve Chu. We knew, in an even more general sense, that "the government" had a role in directing (or supervising, or haggling with) the BP engineering team. But the reality was so much more complicated, and "the government" so much less monolithic. Scores of scientists from the national labs (and the Geological Survey, and NOAA, and even NASA) would do most of the detailed analytical work in the coming weeks—vetting the integrity of subsea hardware, calculating the well's flow rate, interpreting seismic imagery of the geological formation, characterizing the Macondo reservoir, and so on. Chu's science advisers, meanwhile, weighed in from a distance, rarely visiting Houston. They had a special status, and they each brought a unique talent and perspective to the team. They were kind of like the X-Men.

George Cooper had run the petroleum engineering program at Berkeley and had the most relevant experience of anyone on the team. Ray Merewether worked at the Scripps Institution of Oceanography, had extensive knowledge of the marine environment, and could point to a long list of patents for underwater gadgetry. Jonathan Katz, a physicist, had a penchant for the provocative argument. (Bloggers soon discovered that his website included an article titled "In Defense of Homophobia," and Katz in short order came to the realization that his services would no longer be needed.) Alex Slocum, a professor of mechanical engineering at MIT, would soon be the favorite of the X-Men among the BP engineers, because he seemed to have the knack for designing anything at the drop of a hat.

Slocum calls himself a "gizmologist." He didn't show up in Houston to second-guess the BP people, he told me. He preferred to invent solutions to sticky problems. "I was working a whole lot of gizmos. I was having so much fun," Slocum said. He believes that an engineer should look for what he calls "low-hanging rabbits." This is a combination of two concepts: the low-hanging fruit and the fat rabbit. When looking for food, obviously you want the low-hanging fruit. When hunting for rabbits, obviously you want to shoot the fat one. "Why not just get the rabbit and the fruit at the same time?" Slocum said. "Walk in the orchard and look for the rabbit that's on its hind legs trying to nibble at that low-hanging fruit. What are the low-hanging rabbits that we could find?"

Finally, there was Richard Garwin. *The* Richard Garwin, the physicist, the living legend, the 82-year-old protégé of Enrico Fermi.

Now that Steve Chu had gotten the national laboratories, which normally dealt with nuclear weapons, involved in the challenge of plugging the hole in the gulf, perhaps it was only natural that Chu would bring in Garwin, who, back in the early 1950s, had helped invent the hydrogen bomb.

Garwin received his doctorate from the University of Chicago in 1949 and had worked in Fermi's laboratory. A few years later, working under Edward Teller, Garwin contributed to the architectural design of the first hydrogen bomb. That achievement by itself would seem to be enough of a credential to get a fellow in the door in Houston and in the good graces of petroleum engineers, for whom designing a successful oil well might not seem as cool as inventing a thermonuclear warhead. But Garwin's astonishing accomplishments went beyond that. He had played a key role in the discovery of what is known as the Cooley-Tukey FFT algorithm—very complex mathematical stuff, but suffice it to say that he pointed Cooley in Tukey's direction. And the hits kept coming: While at IBM, Garwin helped invent the ink-jet printer. Which is kind of like a miniature oil well, come to think of it.

No one on the Chu team produced more interesting ideas or more sharply expressed opinions than Garwin. He became the team's preeminent thermodynamicist and the most vocal skeptic of many of BP's engineering schemes.

Garwin showed up at the early morning May 12 meeting, took a look at the diagrams of the well, talked to the BP engineers, and began figuring out, with his hydrogen-bomb-developing, inkjet-printer-inventing brain, how to kill Macondo. He had one particularly fine notion: Siphon the oil to the surface from two lines on the blowout preventer, known as the choke and kill lines. Rather than let the oil spew into the sea, the engineers could divert it to a surface vessel and let it wind up in someone's gas tank. By Garwin's back-of-the-envelope calculation, "producing" the oil this way should capture all the flow of the well.

"Producing oil/gas through tubing connected to 3-inch Kill and

Control lines at the base of the BOP would essentially kill the oil flow from the riser into the sea," he emailed his colleagues.

Garwin also had some ideas that didn't really go anywhere.

"Now I'll look at the details of mudding with marbles or with plastic-encase mud sausages injected via the kill and control lines a the base of the BOP. II need to look up the frictional pressure drop in idealized 'porous media' made of marbles or mud sausages," Garwin wrote his fellow team members in an email he probably didn't bother proofreading.

Steve Chu later chimed in by email:

"Building on Dick Garwin's idea of mud balls in sausage casing, what about sending ball bearings in the choke and kill lines before trying a dynamic kill? With several meters of ball bearings at the bottom of the well where the casing pipe comes into contact with the reservoir, an impedance would be created that would decrease the upward flow."

Dick Garwin, again:

"We have mentioned the problem of mudding a vertical flowing well with water-based mud and suggested using sausage-skinned mud balls an inch or more in diameter. Attached is a preliminary analysis of the process, and a suggested garage-scale experiment that uses flowing water and sausage-skinned mud balls. One could start with marbles instead of mud balls, although they would not make an eventual tight seal."

And so on. Interesting ideas! The mud sausages, or mud balls, or marbles, never actually made it from the whiteboard to the deep gulf, but the emails showed brilliant men fully engaged in the Macondo problem. The scientists were *on the job*.

And the BP engineers were just going to have to deal with it.

Early on, Chu's team made a suggestion that gave the scientists a modicum of street cred in Houston. BP wanted to peer into the blowout preventer using high-energy gamma rays. That could give BP a better idea of the condition of the rams that had failed to shear or seal the well. The government scientists had their own gamma ray equipment. At the science team's behest, BP obtained from the Los Alamos National Laboratory a special photographic plate that could improve the gamma ray imagery.

But as the days became weeks, the scientists began to realize that they were being treated as second-class citizens. They weren't always in the room for crucial strategy sessions. BP wasn't telling the scientists what they needed to know. They were given scraps of data. They discovered that a diagram of the well that they had been looking at didn't match the well as actually constructed. Chu told me, "They still thought of us as instrumentation, diagnostic people." They were just the *gamma ray people.*

BP arguably had a valid gripe about the government people. They were as smart as the day is long, but what did they know about Gulf of Mexico petroleum engineering? Chu's scientific advisers could wallpaper an airplane hangar with all their official credentials and honors, but in the eyes of BP, they were a little eccentric. It looked to BP like the administration had said one day, "Where are our experts?" and then rounded up anyone who did not flinch at the sight of a differential equation. Science, engineering, it was all the same. Send in the experts. Did BP really need a Nobel Laureate physicist like Steve Chu barging into the war room? The Obama folks were obviously in love with *the idea* of Chu—this notion of having an in-house Nobel Prize winner who could be dispatched, superhero-like, to solve intractable problems with the power of his giant brain.

And Chu's team? The freelance geniuses? They were an ad hoc addition to an already ad hoc process—ad hoc squared, ad hoc cubed.

BP executive Kent Wells said of the scientists, using his most diplomatic formulation, "As a collective group, they were helpful." But it was not a happy moment for the BP engineers when the big-shot scientists marched in on May 12. The crisis was already several weeks old, and the BP engineers had been working brain-frying hours, doing "parallel processing" to come up with a long list of potential fixes for the problem, trying to get designs from the whiteboard to the machine shops to the dock to the disaster site in a matter of days, when months or years would have been the normal pace of operations. Now the brainiacs were walking in the door saying, Stop everything, we're from the government, and we're here to help.

"There's no way, no matter how smart they are, that they can get up to speed instantly given the pace we were going at. It was an awkward situation at that point," Wells said.

In the War Room

From the very beginning of the crisis, BP's subsea response operation—all the frantic activity in its war room—was largely hidden from public view other than the occasional image or video posted on the company website. Partly this was common sense, as the engineers didn't need to be distracted by journalists and cameras. In addition, BP by general reflex wanted to keep a tight control on its narrative. And there were constant concerns about corporate protection: news, positive or negative, could swing the stock price. At sensitive moments, BP employees would be informed in a company email that the outcome of certain procedures might be "market sensitive" and that they should not say anything that could be construed as a violation of securities laws. One email, for example, stated:

> As required by law, BP is ensuring that all information relating to this operation is publically disclosed (including to the New York Stock Exchange and the London Stock Exchange) on a timely basis. I would like to remind you that you have a duty of confidence and must therefore keep all information as to how the operation is progressing strictly confidential. You must not communicate it to anyone other than as strictly necessary for the procedure. Further, as set out in the Code of Conduct, leaking inside information or tipping someone off can be a breach of insider trading laws which is a serious offense and sanctions can include imprisonment or fines.

BP remained a black box for much of the oil spill response. There were days when the company would be willing to tell you that the sun comes up in the East, but only after running it by the lawyers. There were other days when the company seemed to want to open up a little. Journalists could, for example, get a quick tour of the BP operation in Houston.

The BP war room was not a room at all, but more like a rabbit warren. It had the feeling of a windowless space, though that was largely a trick of blinds and curtains. It was as removed from nature as a casino, a world unto itself, with a security post at the entrance in addition to the one downstairs. It was crowded, with three hundred people working on a given twelve-hour shift. The basic architectural motif of the war room was the temporary table, manned by a person at a laptop that was in turn connected to the rest of the operation by a cable dangling from a hole in the ceiling. The visitor got the distinct impression of hasty improvisation, of a place transformed overnight with wires and duct tape. It was, in fact, already an emergency operations center, one that handled hurricanes and the complex logistics of evacuations of oil platforms and rigs during hurricanes. For the Macondo spill, the place seemed to be geared up to handle the end of the world as we knew it.

The center of the operation, the briefing room, featured an information wall, with printouts, photos, charts, diagrams, all manner of analog material. The wall kept everyone on the same page. At the end of the main hallway loomed a glass wall decorated in a honeycomb design. During sensitive operations, a BP employee would stand guard at the door and require that anyone attempting to enter show a special badge—in addition to the badge that permitted entrance to the war room and the badge that allowed entrance through the turnstiles down on the first floor. This room with the honeycomb design was the highly immersive visualization environment. The Hive. It was the room with the ROV feeds. The Hive was the command center for the delicate subsea operations, all that manipulation of hardware at the bottom of the sea. The technicians in the Hive had a continuous audio link, a kind of twenty-four-hour-a-day conference call, going on with their counterparts on the ships in the gulf. The images from the ROVs dominated a wall of the room.

The first major meeting every day was at 6:30 a.m. sharp. There were typically twenty people in the room. Doug Suttles would be on the phone from the Unified Command center in Robert, Louisiana. An incident commander would go through a checklist: the day's agenda, the safety report (accidents, injuries), the latest engineering progress, the subsea build-out, the dispersants, the oil collected, the oil flared.

The meeting would usually start the moment that Andy Inglis sat down. Inglis, pronounced "*Ing*-ulls," was a Tony Hayward protégé, sitting on the BP board. Born in 1960, he had been with the company thirty years and had risen to the title of chief executive of exploration and production. That covered the planet; the Gulf of Mexico was just one of the ponds in which his operation searched for hydrocarbons. His involvement with the Deepwater Horizon disaster ranged from the very specifics of the subsea response to the more delicate discussions at the corporate level. Sometimes he would have to fly to London. Inglis was the man the other board members would grill about this blasted oil well. Until April 20, he had been on a short list of people who might someday run the company. Now, with the blowout, the short list had gotten shorter by a name—*his* name, and he surely knew it. Inglis would have to kill the well and then see if he still had a job when it was all over. Petroleum extraction is not a famously sentimental business.

Also on hand in the war room, day in and day out, was Kent Wells, who handled the "technical briefings" for BP, which included videos on the company website and teleconferences with journalists. Wells was a godsend for detail-hungry reporters who had been trying to get by on the subsistence diet provided by Doug Suttles in his daily briefings. Wells ran the North American gas business for BP and had been tapped to be Houston's liaison with the news media as well as one of the senior planners. Wells is six feet six inches tall and built like a linebacker. The size, plus an immaculately shaved head, ensures that he will physically dominate a room even if he doesn't open his mouth.

The BP engineers knew that most Americans viewed them as the villains of the story. It bugged them. What bothered them most was that they couldn't do anything to shake their villain status. Killing Macondo would be easier than reforming their reputations. They were going to be the villains even after they succeeded, and they knew it. Wells told me one day that his wife got angry at the media coverage, but he said he tried to ignore it. "If we get at all distracted by that, we won't be as good as we should be on the response," he explained.

The BP engineers were doing what engineers do, routinely, which is devise solutions to problems, but in this case they had to do it around the clock at an astonishing pace. Engineers were on the phone with machine shops dictating how to cut metal on hardware devised minutes earlier. Anyone coming to work had to get quickly up to speed on what was decided in the previous shift. Who could keep track of it all? There was a master schedule. There were team leaders. But in any such complex enterprise, there has to be someone who is completely looped in to every important decision, a person who stays late and arrives early, and puts off any personal needs and issues until the resolution of the Problem.

In this disaster, that was Richard Lynch.

Lynch kept out of the public eye throughout the crisis. What he did, rather than talk to the media or try to soothe the nerves of BP board members, was work, all the time, night and day, seven days a week.

"I did not read the newspapers and did not listen to the news," he told me. "I got to the point where I wasn't even talking to my family anymore" about the response.

When the well blew out, Lynch, a BP vice president and offshore oil drilling veteran based in Houston, was in London on business and could not get back to Houston until April 23 because ash from an Icelandic volcano had grounded flights in Europe. He went straight from the airport to BP's offices for a briefing. He was met outside Building 4 by, of all people, Tony Hayward and Andy Inglis, who had just stepped outside to get some fresh air.

"Have you seen the videos?" Inglis asked him, referring to the images of the burning rig.

No, Lynch answered.

Inglis told him to go inside and get briefed: "You're going to see something you've never seen before."

Lynch supervised the containment work. He was, among the BP engineers, the favorite of Chu's teams, because he was "more in line with our thinking than the hard-core cowboys," as Chu put it in an email to Hunter. Hunter told me that Lynch is "the consummate engineer to pull things together and make things happen." Marcia McNutt said, "I would leave BP headquarters at eleven o'clock at night. I would come back at six o'clock the next morning, and Richard Lynch was still in the same clothes. Obviously he had never gone home. He gave everything to that disaster. He took this very, very personally. This was not just a job."

Lynch told me, "I fish out in the Gulf of Mexico. I could have been on that rig floor. That's the kind of work that I do. This was very personal for us."

And it was also brutally challenging. The industry hadn't adjusted technologically when it crawled down the continental slope to drill in the darkness of the deep.

"The water depth was getting greater and greater, and we felt very comfortable with the equipment that we had," Lynch said. "If I was on land, or if I was in a hundred feet of water, it would have been over in a matter of days."

The engineers in Houston redoubled and retripled and requadrupled their efforts to come up with parallel response plans for containing, sealing, siphoning, plugging, and killing the well. They began sketching out hats, caps, domes. Instead of the containment dome that had been such a big bust, they would attempt to cap the well with what they called the "top hat": a much smaller dome, just four feet across, like an inverted funnel. The smaller surface area, they thought, would mean less hydrate formation, though it would also mean less oil captured.

And they came up with the idea of something called the Riser Insertion Tube Tool. It was a four-inch tube inserted into the open end of the riser right where the major plume of oil was emerging. The RITT didn't have a tight seal, so it wasn't going to end the spill. But BP hoped it would catch most of the flow and siphon it to the surface to a Transocean drill ship, the *Discoverer Enterprise.*

It worked—more or less. The tube went into the riser amid extremely poor visibility. An ROV bumped into a pipe and the tube became dislodged, requiring a do-over. Gradually the operation ramped up and successfully collected some of the flow from the riser. But it could never capture all of the flow, or do anything about the other leak from the kink in the riser above the BOP.

At BP, engineers broke into teams, each tasked with a single tactical maneuver or piece of hardware. Soon there were seven different caps under development.

One line of attack would be to place another blowout preventer on top of the existing one—the BOP-on-BOP solution. Maybe they could, in effect, start over, with new rams and annular preventers and whatnot, and seal the well the way it should have been sealed the night of the blowout.

The BP engineers also introduced the country to the idea of a "junk shot." In a junk shot, technicians would shoot material into the blowout preventer and let it clog, constricting the flow; a dose of mud and then ultimately cement would follow. The material might include scraps of automobile tires and golf balls. BP pointed out that this was all peer-reviewed by the oil industry, that it wasn't something the company had made up on a whim. But when Marcia McNutt looked over the materials being prepared for the junk shot, she was appalled: "It looked to me like someone had taken a neighborhood collection of Happy Meal toys, golf balls, and SuperBalls, and said, 'Let's pump these down the well and see what happens.'"

The Plume

The language of the crisis took on the terminology of a military campaign. The various spokespersons talked of multiple "fronts" in the battle (deep sea, the sea surface, the shoreline). They talked about the "army" of engineers working to design solutions. They talked of the "air force" of planes dumping chemicals to disperse the slick.

But BP was clearly losing the public relations war. The company's brand was fully on the spill by this point, notwithstanding the various attempts to make this the Transocean Deepwater Horizon accident. To the average person following the story, BP caused the spill and, day after day, was failing to fix it.

The spill incited a spasm of national anxiety because it had no obvious boundary. Thad Allen called it right: This was indeterminate. Asymmetrical. Anomalous. The spill did not come from a natural event like a hurricane or an earthquake, but from a technological error. We had to trust that a technological solution would be found for this technological problem—a serious leap of faith given the technological screwup just witnessed.

The situation seemed out of anyone's control. Technology is supposed to do the opposite. "[T]he evolution of the design of machines from the tool stage to powered devices and finally to automation is a process that progressively, and specifically, increases human control," writes Witold Rybczynski in *Taming the Tiger: The Struggle to Control Technology*. This was a case, then, of regressive technology. Machines had created a massive and growing oil slick with no end in sight.

People understood intuitively that this was not just an environmental crisis involving a lot of nasty pollution in the gulf, but rather was a technological crisis. There was a simple yes-no question at the heart of the response to the spill. Did they have the situation under control? Yes or no. And everyone could tell that it was going to be a while before BP managed to get to yes.

As mad as people were, they were about to get madder. On the very day that Steve Chu and his X-Men walked into BP headquarters, the company finally released video images of the leak. The company chose to be extremely parsimonious, showing just two clips totaling about forty-five seconds. One showed the main leak: the big plume of oil and gas billowing from the downed riser. The other showed the containment dome being lowered over the leak and quickly exuding clouds of oil as the dome clogged.

The howls of disgust and outrage at the sight of this plume were accompanied by demands from White House officials, as well as from Representative Ed Markey of Massachusetts, Senator Bill Nelson of Florida, and others on the Hill, that BP release more video. The taste of the plume had created an appetite for more. BP decided to go with full disclosure. It positioned an ROV on the gulf floor where it could stare at the leak around the clock.

So was born the "spillcam." And thus did the people come to know the special, exquisite horror that was the plume.

The thing about this plume was, once it entered the lives of the American people, it never went away. The indefatigable plume shadowed our days and haunted our nights. The plume parked itself somewhere on the screen of every cable TV news program, almost like a network logo. The TV programs would often split the screen, showing an authority figure on the left and the plume of doom on the right. You might see Tony Hayward talking about the company's vigorous efforts to solve this unfortunate Transocean-caused oil spill, and there, right next to him on the screen, refuting his competence, mocking his upper-crust accent, would be the Mephistophelean plume from the depths of hell.

Rule one: No one looks good next to an oil spill.

We didn't always call it a plume. Sometimes we called it a gusher, or a geyser. We said it billowed, spewed, gushed, frothed, boiled, and hemorrhaged. The perceptive Tom Junod wrote in an *Esquire* blog that the English language was unequal to the event. "Blowout" didn't do justice to the event because it did not address the consequences. Calling it a blowout was "like calling the Holocaust an anti-Semitic incident." Taking a cue from Don DeLillo's novel *White Noise* and its central incident, an "airborne toxic event," Junod suggested that "underwater toxic event" would be appropriate for what was happening in the gulf.

My *Washington Post* colleague Hank Stuever found himself unable to stop thinking about the plume: "Spillcam combines the dread of horror films with the monotony of Andy Warhol's eight-hour silent movie of the Empire State Building," Stuever wrote in an essay in the *Post*. "There is no sound and nothing happens, except the inexorable, unending flow . . . You can be in bed and wake up in the middle of the night and think to yourself: It's still coming out. Then you think: What if it never stops?"

On the Huffington Post website, the Reverend Peter M. Wallace wrote of the spill that it's "as though the earth is bleeding for our sins. But in this case, it's not atonement, it's just death."

The plume, or gusher, was not just oil and gas. It was a complex cocktail of oil, gas, water, and sediment. This complicated any attempt to estimate the amount of oil simply from looking at the video. It looked like much more than just five thousand barrels of oil a day—surely, right?—but then few of us knew what five thousand barrels a day might look like. Amateurs did some math. Scientists made calculations. An energetic science reporter at National Public Radio, Richard Harris, emailed the video clip of the plume to three professors, all of whom said it looked like more than 5,000 barrels a day. A Purdue University professor, Steven Wereley, told Harris that, based on his analysis of the brief clip, the well might be spewing 70,000 barrels of fluid a day.

There are natural rhythms to public attention, and to our willingness to engage, meaningfully, emotionally, with a big story. The oil spill refused to comply with our sense that the crisis had gone on long enough. The gusher kept gushing. We were beyond frustrated. We just couldn't stand it for another minute. Why couldn't someone stop it? It was both disgusting and dispiriting. The black plume was a dreadful manifestation of . . . something bad. Our wasteful ways? Our addiction to oil? Our technological hubris, as we ventured into that too-deep water for hydrocarbons? Whatever—the plume was like guilt itself, a constant psychic drain. It suggested that our highly developed civilization is still rather nasty and brutish and perhaps basically incompetent. For the serious hysterics and the full-time prophets of apocalyptic mayhem, the plume served as affirmation of their negativity.

Chapter 7

Louisiana

There was something odd about this oil slick in the gulf. It wouldn't come ashore on schedule. We'd been told, back on April 29, that it would hit land by dusk on April 30, but here it was mid-May, and it was still lurking, huge and unfathomable, with its hidden plumes and variable surface features: sheen, mousse, streamers. The good news was that the Loop Current didn't take it up after all, thanks to a helpful eddy. The oil would not flow to Miami, Palm Beach, Cape Hatteras, Ireland, New Zealand, and Antarctica.

The Mississippi River outlets, opened wide, sent water from half a continent into the gulf, pushing on that oil slick, trying to shoulder it back and keep it from infiltrating the grassy marshes and shellfish-rich bays of coastal Louisiana.

But at last it arrived. The oil, following its own schedule, finally began glopping into the marshes, slathering the beaches of the barrier islands. The oiled birds now became numerous, and so too the oiled turtles. A wildlife rescue operation ramped up not far from Venice, Louisiana. Volunteers dug up turtle eggs on gulf beaches and relocated them to Florida's Atlantic Coast. With the coming of the plume and now the arrival of the oil on shore, the crisis entered a new phase of urgency and repulsiveness.

I flew into New Orleans on the morning of May 16, rented a car, and headed down to Grand Isle, the barrier island with a target on its chest. It's a quirky drive, partially on two-lane roads, partially on highways, with numerous invitations to turn the wrong way. You spend a lot of time hugging the Bayou Lafourche, which, where I'm from, would be called a canal. There's a four-lane road that pops up out of nowhere; it seems to have no name other than "the Four-Lane." You pay a toll at a kiosk and suddenly are flying high above the marsh on an extended elevated highway that zigs and zags for no apparent reason. The world is big here, and watery. This is where the continent seems to evaporate into open water. The sky is big, the horizon distant, the sunsets spectacular.

Almost everything in sight is touched by the hand of human beings. The mouth of the Mississippi used to wander up and down the coast, pumping sediment into the lowlands, but for decades the river has been the ward of the US Army Corps of Engineers, essentially an artificial entity, constantly dredged and imprisoned by levees. Without the supply of sediment from the river, the marshlands—slashed by the oil and gas industry—have been losing the war against salt water. Great chunks of Louisiana have vanished, eaten away as if by a disease. The old-timers would talk of hunting on barrier islands that are now barely sandbars. Nothing out there is as it used to be.

At the end of the road is the fishing village of Grand Isle. The houses are on stilts in deference to the storm surges. It's a one-road town with two water towers, and has the feeling of an old postcard; everything a bit sepia toned. A Starbucks would look as out of place in Grand Isle as a *Washington Post* reporter surely did.

Although I was the last journalist in America to show up to cover the spill, my timing was good, since the oil was just making landfall. This was going to be ugly and tragic, but journalistically it would be easy pickings. With the horror of the spill enveloping this part of the world, everyone and anyone could potentially be a plausible interview subject, all events were portentous, every hour of every day was *the pivotal moment,* and anything that BP might be doing to plug the well was *a make-or-break gambit.* Plus, the people had colorful accents. Fishermen and boat captains and fish market entrepreneurs are good enough, but add to that the seasoning of the Cajun language, and the story writes itself.

During the drive south, aware of the need to file something to update the Web story, I pulled over frequently for grab-and-go interviews with random people, all of them happy to talk, apparently not yet exhausted by the journalistic onslaught. I typed up a feed with an omniscient tone that came perhaps too easily to someone who had been in Louisiana for a grand total of four hours:

> A feeling of imminent calamity continues to pervade Louisiana's coastal towns, where tar balls have been washing up intermittently on beaches and watermen are dreading what they think is the inevitable arrival of the huge oil slick and its penetration into marshes rich in fish, shrimp and crabs.
>
> In Grand Isle, just west of, and tucked underneath, the lengthy claw of the Mississippi River delta, shrimper Harry "Chu Chu" Cheramie, 59, said fishermen are encountering oil not far to the west in Timbalier Bay.
>
> "That's going to kill our fishing grounds. We won't be able to drag that area for a long time to come," Cheramie said.
>
> His wife, Josie, the tourism commissioner of Grand Isle, said people come to the island only for three reasons: "Play on the beach. Fish. Eat the seafood."
>
> Fish market manager Juanita Cheramie—no relation—was fearful on a gloomy and rainy Sunday afternoon.
>
> "We're going to get it. It's only a matter of time. We're just on a wing and a prayer right now," she said. When the oil hits, she said, "It's over. You can lock the gate in Leeville"—a town up the road toward New Orleans.

It wasn't literature, but it was timely, and there were no technical hitches. The AirCard, the little magic wafer that slips into a slot in my laptop, transported the paragraphs straightaway to the newsroom in Washington. The people quoted had interesting names. They possessed a pithy way of speaking.

"You can lock the gate in Leeville."

It was a lyric in a mournful country tune.

Everyone was waiting for the oil. The NOAA maps showed it growing by the day, Rhode Island-sized already, and irregular, patchy, unpredictable. Buggie Vegas, the owner of Grand Isle's Bridgeside Marina (home base of the Labor Day redfish rodeo), told me, "It's like a big hurricane out there, just drifting around. Where it's going, we don't know."

They were all scared, no doubt about it. But where was the anger? In journalism, truth sometimes catches you off guard. Truth can mess up or complicate a good story, interfering with carefully nurtured assumptions.

Sometimes people out there in the field refuse to say what they were supposed to say when, back in the office, we imagined them (just as the oil sometimes refuses to show up on the beach on schedule). In Louisiana, after I'd filed a couple of dispatches and started to listen to people, a new truth emerged: These folks weren't particularly mad at BP. A little bit, sure. But they weren't as mad as the people in Washington. They didn't seem to be saying their anger lines quite as the script dictated. They certainly weren't mad at the oil industry in general.

They liked the oil industry—even now.

Driving into Port Fourchon, I pulled over to talk to Joey Toups, a shrimper. The shrimp season had to be opened early because of the oil spill, and it had been a cold winter, so the shrimp coming off his trawler, *Vamoose,* were tiny, almost one-hundred-count a pound. Two deckhands dumped them into tubs in the back of a pickup for the trip to the seafood processor, an industrial plant somewhere up the road. Toups told me he was a third-generation trawler, but he had worked fourteen years in the oil industry before returning to the waterman's life. He built pipelines and welded for Exxon. Now he does his recreational fishing out by the oil platforms.

"The oil rigs is a reef," Toups said. "It's the best fishing in the world here because we have so many reefs." As he spoke, I could look over his shoulder and see the boosting station of the Louisiana Offshore Oil Port (LOOP). That's where oil from offshore is piped toward inland refineries and storage tanks.

Just down the road, Port Fourchon is a marvel of industry, a boomtown that's all boom and no town. It's just boatyards, warehouses, barracks, temporary office buildings. A truck stop. Cranes everywhere. Port Fourchon really took off in the 1990s when the industry started developing the deepwater fields. It's the closest port to the Deepwater Horizon site. The birthplace of the failed cofferdam has hardly a tree to speak of. Instead of dirt, there's gravel. And yet go a mile in any direction, and there's open water, marsh, nature galore.

The epitome of the small businessman is the guy who gathers oysters from his staked-out oyster bed or trawls for shrimp in a boat on which he's still making payments. The crisis seemed like such an obvious collision of Big Oil and small fish, and yet there was little evidence of populist rage. Maybe it would come in time. Maybe the initial BP claims payments—$2,500 per fisherman, another $2,500 per boat, dispensed in various strip malls and public libraries by BP representatives—had queered the normal level of anger. But most likely, this reflected the long relationship between fishing and oil in southern Louisiana.

"My mama had an oil rig fall on her house," a trawler named Robert Guidry told me as he left an emergency claims office in Galliano. But he was quick to argue that Americans need oil.

"Without that, we'd have to live with candles in our house and go around in a horse and buggy," he said. In this area, he said, "oil naturally comes out of the ground—it's almost like the Beverly Hillbillies."

I met Wilbert Collins, an oysterman still in the business at the age of seventy-two. Collins said there've been lots of spills, just never one as big as the Deepwater Horizon incident. He described what an oyster tastes like when retrieved after an oil spill: "You just taste the oil. It stays in there"—in the oyster—"a couple of weeks."

What worried him more than the oil was the heavy dose of chemical dispersants used on the slick by BP.

"The medicine they're spraying on there, we don't know how bad it is."

Thomas Barrios, a shrimper who had recently opened his own restaurant, chimed in: "Is our kids going to get cancer and all that? Is it going to make people sick?"

I spoke with Martin Folse, the owner of the independent TV station HTV in Houma. He had toured the spill by helicopter, obtaining dramatic footage that he aired on his station. He didn't want to cast any blame for what went wrong.

"I can't sit here and criticize oil because I live in an area that oil has built. But seafood has built it, too. It's two very powerful industries that has been affected at one time," Folse said. "It's like watching two brothers fight. You can't pick a side. You gotta work with both sides."

I pulled over along the Bayou Lafourche to visit with the proprietor of a bait stand, Randy Borne Jr., thirty years old, and having a hard time of it, he told me. The charter boat captains had seen their business take a dive because of the closing of federal waters. No charters, no need for bait. Borne had thought this year would be better than last, when he had suffered due to the recession. He was at the bottom of the monetary food chain and desperately needed help.

He'd heard about the new claim center opening that very morning up in Galliano. The problem was, like many people, he didn't have a lot of paperwork for his business.

"The W-2 form, I sent it off for my food stamps," his wife, Casey, twenty-two, told him.

"You sent off a copy," he said, hoping.

"No," she said.

It got tense for a minute. He stomped off into his trailer home. Casey Borne turned to me and said, "I'm scared that it will come."

The oil, she meant.

"Right now we're not making much as it is."

She went back into the home to confer with her husband. I struck up a conversation with the deckhand on Borne's boat. His name is Clerville Kief III. His grandfather had founded the hardware store in Galliano. Kief lit a cigarette and pondered the calamity.

"You're never going to stop human error," he said. "We don't got nothing against the oil industry around here. We need petroleum products in order for us to operate."

Randy Borne popped out of the trailer and announced the plan. They would show the BP people the trip tickets that documented all of his trips into the marsh to catch shrimp. The two Bornes and Kief jumped in a battered red pickup and were about to leave when Randy Borne got back out and shook my hand.

"Four or five nights, I've had no sleep," he said, and apologized for his abrupt exit.

Two Angry Men

Beyond the continent's edge, the annual dead zone of the gulf, the area with abnormally low oxygen in the water, had already begun to form in the middle of May. Every year, scientists study the massive plankton bloom in the northern gulf where the Mississippi River dumps its fertilizer and organic material. The plankton die, sink, and decompose, a chemical process that depletes the oxygen dramatically.

So how bad would it be when the oil spill collided with the dead zone? It was another unplanned science experiment. "We're in uncharted waters," Nancy Rabalais, executive director of the Louisiana Universities Marine Consortium and the leading researcher on the dead zone, told me.

Uncharted waters, unimaginable events, unprepared responders: Everything in mid-May was un-something. Unprepared. Unlikely. Unbounded. Ungodly.

The oil hadn't hit the beach in Grand Isle yet, but the NOAA map showed it coming ashore any minute at the bird's foot. I took the circuitous trip from Grand Isle to Venice, a journey down the length of the bird's foot peninsula, through Plaquemines Parish. The Mississippi River is arterial in the parish. From the road, Highway 23, you can't see the river, because it's hidden behind the levee, but you do see the disembodied smokestacks of ocean freighters cruising by.

In the village of Empire, Lawrence "Brother" Stipelcovich, eighty-two years old, gave me the local history. He's a lifelong fisherman who can boast that he got the croaker industry going, the pompano industry, the mullet industry. He lives in a house rebuilt since Katrina. The storm erased the village, leaving only concrete foundations.

Stipelcovich walked with me to the top of the levee, which is seventeen feet high. We could see a strong current in the river.

"That's what's saving us: The river's up," he said.

Stipelcovich said that this part of the world used to be radically different, way back before the extraction industries showed up.

"In 1935 the oil companies came down here to Plaquemines Parish. Then we noticed the land started sinking," he said.

Three feet the land subsided, by his calculation. Marshland vanished before his eyes, replaced by open bay. Where his father once dug a narrow canal for boats to reach a nearby bay, there is now wide-open water.

At the end of the road is the fishing village of Venice, a collection of marinas and boatyards. The occasional alligator patrols the roadside swamps.

Seabirds and herons stalk the culverts and canals. It's a transitional place that doesn't feel like "land," exactly. One senses that Brother Stipelcovich was right, that the land has sunk, because the tidewater comes right to the edge of the two-lane road, and if you go far enough, the road becomes entirely submerged.

This remote village had new inhabitants, a buzzing swarm of sheriff's deputies, Coast Guard officials, reporters, camera crews, and two public officials. Louisiana governor Bobby Jindal and Plaquemines Parish president Billy Nungesser took turns in front of a bank of microphones. The two men made an arresting pair. Jindal is an Indian American, small, wiry, a sparkplug of a man, while Nungesser is a good ol' boy, colossal compared to Jindal. Nungesser is built like a cofferdam.

Here, finally, was the anger. Jindal and Nungesser were angry in different ways, or with different styles of demonstrating their displeasure. Jindal's was an anger fed by urgency, impatience, frustration. He spoke very fast, almost frenetically. Nungesser was just mad as hell. Nungesser looked like a man who'd been hot under the collar since the day he learned to walk.

Jindal had enjoyed a good spill, politically. He seemed to be fully engaged in every detail of the state's disaster response, able to speak without notes about obscure Louisiana agencies that he'd directed to do this and that. Hardly a day passed that he didn't jump on a helicopter and survey the encroaching oil slick. He had clearly learned the lesson of Katrina, that politicians in a crisis need to roll up their sleeves and look every bit as alarmed as the citizenry.

Jindal and Nungesser had a plan, stunningly ambitious but environmentally suspect. They wanted to launch an emergency civil engineering project that would cost hundreds of millions of dollars. The idea was to create some islands, in a chain, along the Louisiana coast, and let these islands catch the oil, the gunk, the tar balls, and so on, before the oil could reach the fragile marshes. They would create the new barrier islands, or sand berms, using heavy dredges that would scoop up the sand in shallow water.

No one could accuse them of doing nothing. They would create emergency islands from scratch.

Jindal, speaking to the media, compared the situation to cancer.

"We know we've got cancer. We know we need treatment. We know we've already had some damage. We have to stop this cancer from spreading," the governor said.

Nungesser, when he came to the microphone, complained that the federal government had not yet given permission for the sand berms.

"It's gotta happen. How can they say no?" Nungesser asked. He said he had begged the Army Corps of Engineers to grant an emergency permit to start dredging. He said he'd been in conference calls with federal officials and environmental organizations. He said the rush-job nature of the project was contrary to the usual pace of government action.

"They want to know how the *snail* is going to be affected in Florida," he said.

This was a milestone day: Oil had been detected in the marshes. It wasn't offshore anymore. The gunk had hit the canes of grass. Nungesser said this was the nightmare he'd been fearing.

"Everything that blanket of oil is covering today will die," he said. "There's no way to clean it up."

He went on: "We lost a battle today. We've got to win this war. Failure is not an option. Where we're standing today—there will be no marina, there will be no fishing. If this continues, everything will be dead for many years."

Later, in the marina restaurant, he looked over photographs of the spill on a reporter's laptop.

"Look at that. God help us."

The only problem with the Nungesser-Jindal sand berms was that every federal agency that looked at the plan recoiled in horror. First of all, it didn't make any sense as a response to an oil spill. The sand berms would take six to nine months to create, minimum. The oil was coming ashore *right now*. There might be good reason to dredge and rebuild barrier islands as part of a comprehensive and carefully designed coastal restoration project—the Louisiana officials had been pushing for such a project for many years—but it wasn't something you'd want to do as a rush job without knowing how it would affect currents, natural transport of sand, erosion rates, salt water levels in the bays, marine life, and on and on.

The federal agencies wanted nothing to do with this. Thad Allen, though, was going to hear from these Louisiana officials every day. They were insistent: Sand berms! Sand berms! Sand berms!

The Investigators

As the oiled birds and dying turtles led the news reports, BP kept scrambling to contain, disperse, skim, or burn the oil even as it ran through all the possible options for plugging the well. The company found itself buried in suggestions from mechanics, engineers, oil industry gadflies, inventors, tinkerers, geniuses, and just about anyone who'd ever had a plumbing problem. The correspondents trembled with their knowledge of how this well could be plugged if only BP would set its mind to it.

By mid-May, BP had already gotten 15,000 suggestions, and the government had received thousands more. Most involved some kind of cap or plug on the well that would instantly shut off the flow. It was, after all, just a hole at the bottom of the sea. Why not, say, drop a giant lead ball down its gullet? Why not pump cryogenically chilled –90°C methanol down the well make the oil solidify—*ooph!*—just like that, an instant clog? Why not cut off the damaged riser and attach a new thingamajig that you could just crank closed?

BP, still under the impression that it might position itself as the good corporate citizen, assured the public that the ideas were being taken seriously by the company, though they were no more likely to wind up in BP's battle plan than become enshrined in the Constitution. BP had enough ideas in-house. And there was a common problem, as BP saw it in those early weeks, with many suggestions from the general public: they assumed a well with integrity, a well that could handle the spike in pressure if the flow were suddenly shut off.

BP didn't feel confident about that. Nor did the government scientists looking over the shoulders of the BP engineers. They were all worried about making the situation worse. The real fear was the underground blowout.

The anxiety was incited by the design of the well. Macondo was designed to be a bit . . . cheap. "Minimalist" is the word that Richard Sears, the veteran deepwater oil expert who worked for the Oil Spill

Commission, prefers. "You had this sort of overarching philosophy: 'I can do this in a minimalist fashion,'" he said.

There just wasn't much steel down there, not much pipe. BP designed the well with a series of telescoping casing strings, getting narrower as the well went deeper, each casing hanging from the wider casing above it. The final casing string was the "long string," more than two and a half miles long and running from just below the wellhead to the bottom of the well (with a little bit of open hole below that, an area known as the "rat hole"). The long string measured 9⅞ inches at top and 7 inches at the bottom.

In between that "tapered" casing string and the outer casings higher up was an "annular" space, also known as the annulus. It was kept filled with mud. Along the 16-inch casing were three "rupture disks." These were emergency vents of sorts that enabled the well to relieve itself of pressure in a pinch. During the normal production of oil, the central casing would be filled with hot, flowing hydrocarbons. The adjacent annulus would heat up. Heat raises pressure. What worries an engineer is that this confined annular space could potentially build up a dangerous amount of pressure. The pressure could conceivably collapse the inner production casing and ruin the well. So the rupture disks allowed pressure to escape into the surrounding formation. The rupture disks were designed to work if the annular pressure reached 7,500 psi.

But this was an unfortunate design in light of the blowout. The engineers wanted to kill the well, but the rupture disks complicated their task. What if the annular space was where the oil and gas were now flowing from the reservoir?

They said to themselves: Let's assume we cap the well. Put a seal on it, or a new valve of some kind, or an entirely new blowout preventer atop the old, defective one. We shut it in. We turn off the well just like turning off a faucet in the backyard. What happens? You've got a column of oil and gas surging from a reservoir, and maybe it's coming up the annulus. With the new cap on the well, the annular pressure builds. And then: *foom!* The oil and gas surge out the rupture disks and into the crumbly rock formations below the seafloor. Then keep surging. The hydrocarbons wend their devious way higher and higher in the formations, lured by the lower pressures in the shallower rock. Eventually the oil and gas pop into the bottom of the gulf. There are now five leaks, ten leaks, fifty leaks, it's the full nightmare, you've got fissures growing larger, eroding, you've got cratering, you've got gushers shooting out the gulf floor, and you know you'll never be able to play whack-a-mole fast enough to kill all the leaks.

So that gave everyone pause.

Such a scenario could potentially destroy the chances of the relief wells to kill Macondo. The relief-well operation was premised on the ability to apply some controls to the flow of the well at the wellhead, perhaps via the largely defective blowout preventer or some other apparatus, even as the second well comes in and shoots mud and cement into the bottom. If there are 300 leaks, you've lost any ability to manage the well even in theory.

The upshot was that, as engineers devised plans A, B, C, D, E, F, and so on to kill the well, they also pondered a mystery: What was happening down there? Where was the flow? Had the initial blowout damaged the well? Had it blown through the rupture disks already?

And that, in turn, brought up the initial question, the one that the Horizon's Houston-based rig manager, Paul Johnson, had asked Jimmy Harrell right after the explosion: What happened?

For the people who knew offshore drilling, who spoke Engineer, who knew their mud weights and pore pressures and fracture gradients, who understood cement bond logs and wireline logs, who routinely discussed negative tests and positive tests and seal-assembly tests, there were lots of questions that needed to be asked.

Why had the cement job failed?

Why hadn't the negative test on the cement job revealed the hydrocarbon invasion of the well?

Why didn't the rig workers notice that something was whack?

Did the VIP visitors distract the rig crew and rig managers when they visited that afternoon?

Why did the hydrocarbons (the engineers would say "HCs") get all the way up into the riser and then onto the rig itself before anyone tried to disconnect the rig from the well?

How did the gas find an ignition source on a rig that's designed not to produce sparks of any kind?

Why didn't the blowout preventer seal the well?

And, finally: Was the oil and gas coming up the central casing or up the annulus?

People had to know why Macondo did what it did, because the solution to this horror might ultimately pivot on its cause.

In the early weeks, amid the confusion of the disaster, there would be many explanations of what happened, most of them wrong, or at least wildly oversimplified, as if demonstrating a human instinct for reductionism, order, logic, this-follows-that. People prefer Newton's universe to Einstein's. They like the idea that one thing—cue ball hits eight ball—leads incontrovertibly to the next: eight ball in corner pocket. They would like history to be sufficiently deterministic that, if the tape were replayed, things would turn out pretty much the same way. And they like a universe of sufficient morality that bad things happen most often to bad people.

Ideally, this Deepwater Horizon disaster could be explained by a single, glaring sin, such as greed. There is some of that lurking in the story line, no doubt. If not greed, then greed's sniveling, mouth-breathing cousin, cost consciousness.

People wanted a single technological explanation for what happened. A congressional aide told me, on background, not for attribution, hush-hush, that, based on briefings his boss had received, "The best theory would be that there was some type of problem with the cement."

The cement! That had the ring of truth, the ping of significance. Bad cement leads to oil spill. Some of the left-leaning bloggers loved this explanation: If it was Halliburton at fault, that meant that the person really responsible for all this was . . . Dick Cheney. Who had run Halliburton in the 1990s.

There was just one niggling detail that made the cement explanation inadequate: It was not really an explanation at all. Because why did the cement job fail? Wasn't the cement tested in the lab, tested in the field, by expert drillers on a blue-ribbon rig? And what about that negative pressure test?

Macondo passed that test on the kill line, didn't it?

The other easy explanation: bad blowout preventer. The early, glib response to what went wrong would cite this bad piece of hardware. From the BP perspective, blaming the blowout preventer for the problem had the nifty feature of shoving to Transocean the primary culpability for the Transocean Deepwater Horizon blowout, since Transocean owned, operated, and maintained the BOP.

Except that a blowout preventer can't cause a blowout. It can only strangle the blowout after it has already started. A blowout preventer is a backup plan, a fail-safe. The Macondo blowout was really a double whammy: a loss of well control deep in the well (bad cement job, whatever), followed by a failure of the blowout preventer—and perhaps, just as important, the inability of human beings on the rig—to stop the explosion and/or shut down the flow of gas that was feeding the fire.

Blowouts cannot result from a single failure, because in any drilling operation there are safeguards, redundancies, tests, protocols, reviews, real-time well monitoring. The well passed the positive pressure test, the seal assembly test, and the negative pressure test. It looked good. The well had full returns of drilling mud, like it was supposed to. The cement landed where it was supposed to land, by all accounts. All signs were benign.

So something here was goofy, in a big way. When investigating a disaster like this, one has to probe an entire engineering system. Why did the system so crammed with redundancies and safeguards fail? Why didn't the people involved, who are trained to perceive hazards, realize they were minutes from an explosion? What was it about the design of this well that could have led to the catastrophic failure? Was there something within the culture of the organization, something innate to the corporate psychology, that permitted the calamity?

In the case of the Deepwater Horizon, many things had gone wrong, in succession, in a cascade of error and misfortune. So much had gone wrong that it's almost hard to know where to start.

The disaster essentially created a new industry, the Deepwater Horizon Disaster Investigation Industry. There were multiple inquiries operating in parallel, with differing levels of public attention, legal muscle, technical expertise, and objectivity.

Some of these investigations had to go far beyond figuring out why Macondo blew. They had to decide whether Macondo was a one-off event by a rogue operator—a freak occurrence, a Black Swan—or part of a more systemic problem in the industry. Was this a BP problem or an oil industry problem or a government regulation problem or a deepwater drilling problem? If this was specific to the Macondo job, how would the culpability be apportioned among BP, Transocean, Halliburton, and the other, more obscure contractors, such as M-I Swaco, Weatherford International, and Dril-Quip?

One investigation boasted the broadest portfolio, the highest public profile, and the weakest legal power. It was the commission appointed by President Obama, officially called the National Commission on the BP Deepwater Horizon Oil Spill and Offshore Drilling. The administration appointed as cochairs of the panel two estimable Washington figures: former senator Bob Graham, a Democrat, and William Reilly, a Republican who had directed the Environmental Protection Agency during the George H. W. Bush administration. Although the House of Representatives voted overwhelmingly, in bipartisan fashion, to give the commission subpoena power, a cluster of Republican senators objected, and the commission was forced to do its work with no more power than a pleading smile. Given that limitation, the commission did remarkable work, led by some young, crackerjack lawyers with engineering backgrounds and some veteran engineers from the oil patch.

A number of congressional committees launched their own probes. Representative Henry Waxman, the California Democrat, put his Energy and Commerce Committee investigators on the case. Michigan Democrat Bart Stupak, with the Oversight Subcommittee, had a piece of the action. So did Ed Markey, whose Subcommittee on Energy and the Environment would lead the charge against BP for obfuscating on the flow rate.

The Justice Department got involved, though its investigation would prove to be rather inscrutable as the weeks went on. Although the furious public would be thrilled to see BP executives in handcuffs, there was the nagging question of what, exactly, the specific crime might be. There is a point at which an accident is sufficiently foreseeable that one enters the realm of gross negligence and criminality. But what Justice might be thinking was an ongoing mystery.

There were some less publicized investigations, too, including one by the US Chemical Safety Board, a congressionally funded organization with a sterling reputation for inquiries into industrial accidents. The Chemical Safety Board knew BP already, having probed the explosion at a BP refinery in Texas City, Texas, in 2005, that killed fifteen people. Also looking into the accident were panels from the National Academy of Engineering and the American Petroleum Institute.

Of all the investigations, one proved to be the most consequential, at least in terms of producing sworn testimony about what happened. This was the Marine Board of Investigation, a joint effort of the Coast Guard and the Minerals Management Service. It was a very odd affair from the get-go. The Marine Board was a fact-finding investigation that would culminate in a report, and potentially a recommendation for action. It might, for example, recommend changes in government policy. It could push for a sanction against one of the people or companies involved. But it did not, by itself, make rulings. It did not prosecute. It held hearings every month or so, not in a courtroom or lofty government office building but in the ballrooms of modest hotels close to airports and freeways. The Marine Board hearings featured testimony from survivors of the blowout; the companies involved made sure their best lawyers were on hand, which meant that well-heeled barristers who bill $950 an hour were crammed shoulder to shoulder from one side of the ballroom to the other.

One other, important investigation was led by BP itself and headed by Mark Bly, the company executive in charge of safety. Bly put together a team of fifty people, some from outside the company. Though obviously not disinterested, the Bly team considered itself an independent operation, working behind locked doors in a suite of offices at Westlake Park headquarters.

Seven days after the blowout, the Bly team questioned Donald Vidrine, the company man who made the crucial decision to continue pulling mud out of the well. The interviewers took notes, sometimes referring to Vidrine in the third person, sometimes paraphrasing him in the first person. This was not a legal document, no one was under oath, and Vidrine never signed off on the notes—so this was not exactly like testimony, and it would never be admissible in court (nor easy to understand if you don't know the industry lingo). But it became part of everyone's narrative. Bly committee notes made the rounds.

An excerpt of the notes:

> *Cement job went ok, complex job with different fluids, looked like it went fine, bumped the plug and tested the seal assembly.*
>
> *Pulled out of the hole to pick up the cement stinger.*
>
> *Ran in hole to the top of the BOPs and tested the casing, very good test based on the chart he saw.*
>
> *Carried on running in the hole to depth, rigged up and started the displacement. Had a procedure, displaced the choke and kill lines, pumped the displacement pill followed by water . . .*
>
> *When he arrived, there was 1,400 psi on the [drill pipe]. He said they shut the DP back in and it had come back up. He questioned this and was told by the team in the drillers doghouse that this was annular compression—he had heard about this but had not seen it before.*
>
> *He went to the kill line and there was no pressure . . .*
>
> *The 1,400 psi remained on the drill pipe, it stayed at that pressure—there was no indication that it was gas coming up . . . Told them to let me know when the pill comes up so that we could do a sheen test.*
>
> *I then went to the office and checked what calls I had.*
>
> *They called to say the pill was back so I went to the rig floor.*
>
> *Everything was fine, checked for flow and it was ok . . .*
>
> *The sheen test was passing . . .*
>
> *Said to go ahead and start dumping . . .*

For most of us, these notes are in code and are utterly incomprehensible. But what jumps out of Vidrine's comments (as translated/interpreted by the investigators) is that this veteran company man—someone who had spent many years of his life offshore—did not seem particularly alarmed. He'd been worried earlier about the negative test, and discussed it at great length with the drill team, but his actions were not those of someone oppressed by a sense of imminent catastrophe.

When investigators asked Vidrine what he thought went wrong, he answered, "I have no idea!"

Which echoed the Jimmy Harrell line: *It just blew.*

The Bly team interviewed Brian Morel, an engineer who normally worked in Houston on the Macondo well team but had spent six days and five nights on the rig, leaving on a helo at eleven o'clock on the day of the blowout. He was familiar with every detail of the endgame on the well. He wasn't the only engineer involved, but he was definitely up to his elbows in the final casing run and the cement job and all the procedures that the Houston team had developed.

Morel told the investigators that the rig was safety focused. An investigator jotted some notes:

Rig is perf driven

One step ahead always

Great wells over 9 yrs

Touted one of best Transoc & BP

Every Mtg talk about safety

Everyone first safety

Morel's account echoed what they had heard, and would hear, elsewhere: This was a great rig, great workers, it had drilled great wells, and they were all safety-conscious.

And yet their well exploded.

Morel ran through the problems with the drilling of Macondo, including the narrow gap between the pore pressure and the fracture gradient, how they had to calibrate the mud weight very precisely, how it was like threading a needle down there in a realm they couldn't see directly.

Mud too heavy, it busts up the rock and leaks away; mud too light, it's bullied by the hydrocarbons.

In a separate list, Morel detailed fifty-nine distinct steps involved in the setting of the final production casing in the well and the cement job. The list of fifty-nine actions made clear that the temporary abandonment of a well is not a simple operation. Many of the procedures are routine, however. Anyone reading the list would have to squint to try to figure out where the misstep had occurred. The final eleven steps had an almost monotonous quality until finally . . . boom:

Stung back into seal assembly and tested again to 10,000 psi for 10 seconds followed by 6,500 psi for 5–10 min (straight line 4–5 psi/min drop—good test)

Monitor well on trip tank/no flow

Pick up to 4,770 and dropped nerf ball/circulated 1–1/2 DS volume

Pumped 30 bbl slug at 16.3 ppg

POOH to surface

Test casing per APD to 250/2,500 psi

RIH to 8367'

Displace to seawater from 8367' to above the wellhead

With seawater in the kill, closed annular and did a negative test ~2,350 psi differential

Open annular and continue displacement

Blowout Occurred

How to break that code? Somewhere in there was a solution to the mystery of the blowout.

Chapter 8

Top Kill

BP's Bly team investigators combed through company emails, which were emphatically not the kind of emails a large corporation wants to unearth in the wake of an environmental catastrophe that has shut down a large portion of a region's economy and rankled the president of the world's most powerful nation and so forth. The emails showed engineers changing plans on the fly, fretting and improvising. They seemed to have time pressures, and a keen awareness of how much the Macondo job was costing the company.

On April 9, David Sims—one of the two BP executives who would visit the rig eleven days later—sent an email to colleagues outlining the scheduling dilemma facing BP. The company had approval from MMS to use the Deepwater Horizon for two projects after Macondo, one at a well named Nile, the other at a well named Kaskida. But the delays in the Macondo job threatened to shred a very tight schedule. The mood within the well team was captured in an email on April 13 from a BP engineer, Merrick Kelley, to engineer Brian Morel:

"I know you all are under pressure to finish Macondo so we can get Nile P&A moving and not jeopardize the Kaskida well . . ."

On the fourteenth, Morel wrote his colleague Richard Miller about some possible last-minute changes in well design:

"Sorry for the last notice, this has been a nightmare well which has everyone all over the place."

More problematic were the emails about the cement job on Macondo. On April 15 Jesse Gagliano, a Halliburton employee embedded in the BP well team in BP's headquarters, sent the BP engineers an email saying that a computer model of the cement job, as designed, "shows the cement channeling." Gagliano on April 18 emailed a thirty-page report to the BP well team showing that, as designed, the well had the potential for a "*severe* gas flow" problem.

Channeling means a bad cement job: gaps in the cement that can let gas infiltrate the well. That could lead to a gas kick or a blowout. Channeling is not necessarily catastrophic, but when you're an investigator trying to find out why the well blew up, and you see an email written just five days earlier saying "shows the cement channeling" and another report saying there could be a "*severe* gas flow" problem, you are hearing alarms and sirens going off.

Gagliano's computer model, called the "OptiCem" model, indicated that the channeling hazard would go away if the final production casing were centered in the hole with twenty-one "centralizers": sleeves that fit over the pipe and keep it positioned equidistantly from the sides of the hole. The rig had only six centralizers on board, however, as Brian Morel pointed out to Gagliano. As far as making any changes, Morel wrote, "[I]t's too late to get any more product to the rig, our only options is to rearrange placement of these centralizers."

BP's Macondo well team in Houston worked essentially side by side with Gagliano. He was supposed to move on to a different Halliburton position soon, and it wouldn't be soon enough for the BP people. Morel complained that Gagliano was late in producing test results on the stability of the cement slurry recipe ("Jesse isn't cutting it anymore," Morel wrote a colleague). They were about to cement a well with a concoction of cement that they couldn't know for certain would be stable in the hot, deep, pressurized well. Gregory Walz, a BP engineer, discussed the centralizers issue with David Sims, the boss ("He agreed that we needed to be consistent with honoring the model") and, late on April 15, the BP team arranged to send fifteen centralizers to the Horizon on the next flight out, in the morning. They also lined up an employee of the drilling contractor Weatherford who could install the centralizers.

But then that plan fell apart. At midday Friday, the sixteenth, John Guide, the well team leader for BP in Houston, wrote to Walz that he had just learned that the centralizers were the wrong kind and had too many distinct parts that might potentially come off the casing and jam the operation at a crucial moment. "Also it will take 10 hrs to install them. We are adding 45 pieces that can come off as a last minute addition. I do not like this and as David [Sims] approved in my absence, I did not question but now I very concerned about using them."

Guide and his colleagues decided to go with the six centralizers originally on the rig, spread them out across the hydrocarbon zones, and hope that that would keep the pipe adequately centered in the hole. If they didn't have evidence of a good cement job, they could run a cement bond log test, look for areas that had channeled, and do a "squeeze job," sending down more cement to the area that needed reinforcements.

On Friday, April 16, BP engineer Brett Cocales wrote to Guide:

> Even if the hole is perfectly straight, a straight piece of pipe even in tension will not seek the perfect center of the hole unless it has something to centralize it.
>
> But, who cares, it's done, end of story, will probably be fine, and we'll get a good cement job. I would rather have to squeeze than get stuck above the [wellhead]. So Guide is right on the risk/reward equation.

What a disaster this email was. This wasn't technically a smoking gun (BP investigators came to believe that the centralizers were irrelevant to the blowout), but it would read like one in the media. "Who cares, it's done, end of story"? It could hardly have been worse. And yet, read closely, in the context of the other emails, Cocales was not being cavalier at all. The well team had to decide whether to go with centralizers that they believed had forty-five separate moving parts, and which could potentially bollix up the whole operation, or go with just the six spread-out centralizers and run a risk of an inadequate cement job that might be remediated with more cement. What's clear in Cocales's memo is that he was not considering the possibility of a blowout. His worst-case scenario was having to conduct a squeeze job ("I would rather have to squeeze") to fill in the gaps in the cement. And a squeeze job isn't a disaster, it's just a money issue, because it takes time and delays the operation.

Words are like overpressurized reservoirs: They can explode. Words can go rogue, off the reservation, gamboling about the media landscape, pillaging and plundering—a mess of mixed metaphors being only one possible nightmare. All that anyone would remember of the April 16 Cocales email was that awful verbal shrug:

Who cares, it's done, end of story, will probably be fine . . .

"That's Why We Have Them Pinchers"

The Marine Board of Investigation, the joint inquiry of the Coast Guard and the Minerals Management Service, began its second session of hearings on May 26 in Kenner, Louisiana, in the ballroom of the Radisson Hotel not far from the New Orleans airport. Among those in attendance were personal-injury lawyers who represented rig workers and could use the testimony in the hearings as "free discovery"—material that could be used subsequently in civil lawsuits against BP, Transocean, and the other companies involved. The members of the Marine Board were not lawyers, and so the proceedings sometimes had an amateur-hour feel to them, degenerating into acrimony.

The board's cochairman, Coast Guard captain Hung Nguyen, had a particular focus on the chain of command on the Deepwater Horizon. He seemed to feel that there was a dysfunctional command structure, one in which the command could shift instantly, in an emergency, from one person (Jimmy Harrell, the offshore installation manager) to another (Captain Curt Kuchta). That made no sense to Nguyen. He frequently peppered witnesses about Kuchta's actions. The captain's attorney, Kyle Schonekas, became convinced that Nguyen wanted to yank his client's license.

The other cochairman was a gruff federal regulator, David Dykes of the MMS. It did not escape public notice that the MMS was investigating an incident in which its own alleged failure to regulate the industry properly would be a recurring theme. The MMS was seriously tarnished by the oil spill, but it had problems even before April 20. It had been eviscerated in a recent internal government investigation that detailed a too-cozy relationship between MMS regulators and industry executives.

The agency had an inherent conflict of interest, because it collected oil royalties on behalf of the government while regulating oil practices. The Obama administration eventually declared that the name "Minerals Management Service" would never be heard again. The agency was restructured and renamed the Bureau of Ocean Energy Management, Regulation and Enforcement (BOEMRE). The name was a bit clunky and bureaucratic, but it also provided a snapshot of the political moment: The government would, from now on, manage, regulate, enforce, investigate, punish, and take no guff from the offshore drilling industry, rather than lick the oil off the industry's boots. Just from the new name of this agency, you could tell that the regulators would not be the type of people who would ever crack a smile.

All that political backdrop aside, the Marine Board had an investigation to conduct. One of the first witnesses of the second session was Douglas Brown, the chief mechanic for Transocean, and he wasted no time in dropping a bombshell. Brown described a midday meeting on the rig on April 20, less than twelve hours before the explosion:

> I recall a scrimmage taking place between the company man, the OIM and the toolpusher and driller concerning the events of the day. The driller was outlining what was going to be taking place. Whereupon, the company man stood up and said, "No, we have some changes to that." I really didn't pay attention to what he was saying. They had to do with displacing the riser for later on that tour. And the OIM and the driller and the toolpusher had a disagreement with that. There was a—I remember there was a slight argument that took place and a difference of opinions, and the company man was basically saying, "Well, this is how it's going to be," and the toolpusher and the OIM reluctantly agreed.

The protocol of the hearings allowed the investigators on the board to ask questions first, followed by a lawyer representing the "flag state," the Marshall Islands, followed by the attorney for the witness, followed by the private counsel for the various parties in interest. Brown's Houston-based attorney, Steve Gordon—who specializes in personal-injury cases involving offshore drilling—made good use of his turn, eliciting a key detail that Brown had neglected or forgotten to bring up.

> Q: Do you recall him after the interaction between the BP person and the OIM, do you recall Jimmy Harrell as he was walking out saying anything?
>
> A: Yes. He was—
>
> Q: What did he say and how did he say it?
>
> A: He pretty much grumbled in his manner about, "Well, I guess that's what we have those pinchers for."
>
> Q: Repeat that.
>
> A: "I guess that's what we have those pinchers for."
>
> Q: Okay. Do you know what he was referring to or why he would have said that?
>
> A: I'm assuming he was referring to the shear rams on the BOP.
>
> Q: Okay. Thank you so much.

It sounded like Jimmy Harrell had all but predicted a blowout. The pinchers are the blind shear rams on the blowout preventer. They're hydraulically powered rams, sharp and strong, that slice through the drill pipe from opposite directions and completely shut in the well.

The reasonable observer would take away from Brown's testimony the understanding that the BP company man (Robert Kaluza) forced the Transocean guys to do something out of whack, and then pulled rank ("this is how it's going to be") when people pushed back.

This would be the first of many "aha!" moments during the course of the Marine Board investigation. Indeed, there would be so many such moments that they would crash into one another, a fifteen-car pileup of revelations, telling facts, pivotal details.

Kaluza would not testify before the board—via his lawyer, he invoked his Fifth Amendment right against self-incrimination. Vidrine, the other company man, also would not testify, citing medical reasons. But many other people would take the stand, some of them with intimate knowledge of what happened April 20 on the rig, some of them offering a vivid look at the thinking of the BP well team in Houston, and some contradicting the testimony that had preceded them. Among the most important witnesses was Jimmy Harrell, who appeared before the board on May 27, the day after Douglas Brown.

Harrell said there was no "debate" with Kaluza the day of the explosion, but rather a discussion about the drilling plan of the day.

"We did talk about the negative test. He had given me a plan, and I looked at it, and it didn't have anything about a negative test. We just remained after the meeting, and I talked with them and the driller and the senior pusher, you know, to make sure we did a negative test before displacing the seawater," Harrell testified.

As for the "pinchers" comment, Harrell said that happened on the previous day, the nineteenth of April, and came after he learned that BP was going to use nitrogen-foamed cement, something that worried him because he'd had some problems with nitrogen in previous cementing jobs.

Most significant, Harrell said he had no inkling that the rig was in danger of experiencing a blowout. He was in the shower when the explosion happened.

MMS investigator Jason Mathews quizzed Harrell about whether he or anyone on the rig felt rushed to finish Macondo.

> Q: Was there any pressure on you from anyone within Transocean or BP to complete the job that you were doing at Mississippi Canyon 252?
>
> A: Not at all.
>
> Q: What is the day rate of the Horizon, sir?
>
> A: Five hundred twenty-five a day.
>
> Q: So there was no pressure, from what you told me, being about twenty million behind on a well?
>
> A: No, sir. Not on me, for sure, or my people.
>
> Q: Is there any time you feel you have operational safety being affected by rig efficiency and rig rates?

A: I'm sure at times people want to get it done and try to meet timelines, but no, you never jeopardize safety.

Mathews quizzed Harrell about the history of problems during the Macondo job. Harrell didn't make those problems seem like a huge deal. He summarized weeks of trouble in just a few phrases.

Q: Can you please elaborate on what kinds of problems you encountered?

A: It was several cases of lost returns, and we took a couple of kicks and had stuck pipe. Actually had a side attraction.

Q: Can you please inform me of what a lost return is.

A: That is when the formation won't hold the weight of the drilling fluid. The hydrostatic pressure exerts.

Q: How about a kick?

A: That is when your mud grade is not dense enough to handle the pressure exerted by the formation.

Q: And all of these are potential problems that could lead to later problems in the well?

A: Well, I wouldn't say that, you know.

Now came the questions about the negative pressure test, the key event in the hours before the blowout. Although BP's company men, Kaluza and Vidrine, had the ultimate authority on interpreting the test, Harrell and his Transocean workers were very much in the discussions and could have objected if they did not think it was safe to continue displacing mud from the well.

Q: How many times did it take to successfully test the production casing cement job?

A: Negative, positive, or what?

Q: Negative.

A: They had a good test the first time, but they wanted to do it again before the company man come on, Don Vidrine.

Q: Why did they want to do it again?

A: They wanted to bring it up the kill line also.

Q: Was there any discussion between BP and Transocean on this procedure or why they were doing this?

A: No, sir. I mean, I don't have any—I don't have any problem, you know, if they want to do another negative test.

A lawyer for Halliburton, Don Godwin, pressed Harrell about what Douglas Brown had referred to as the "scrimmage" with the BP company man.

A: [I]t wasn't no argument. I did ask him to stay behind. Like I said, all these plans kept changing, and the plan I seen did not have a negative test to be performed before displacing with seawater.

Q: And the plans were BP's?

A: Yes. They had been coming from town. They had been changing pretty regular.

Q: Did it concern you that BP was regularly changing its drilling plans on the Deepwater Horizon certainly in the immediate days prior to the blowout?

A. It didn't concern me. A lot of it was based on they were scared they were going to lose returns.

Godwin asked Harrell about the pinchers comment.

Q: Did you make the statement following the meeting on the nineteenth, when you left, that "that's the reason we have the pincers"?

A: I can't recall.

Q: You are not saying you didn't say it, you just don't recall it?

A: I'm not saying I didn't. But I just don't recall it.

Steve Gordon, Brown's lawyer, made another run at the pinchers remark.

Q: Regarding the pinchers, sir, I want to make it clear what you said. Is it true that you said, "Well, that is what they make those pinchers for"?

A: I probably did say that.

Q: Looking back on it now, can you identify any warning signs at any point in the day of the explosion that would have suggest a potential event like this was going to happen?

A: No, I couldn't. If I was aware of any problem, I would have been on the rig floor to resolve any problem that I thought may have occurred.

Harrell's testimony was rather confounding. He had more or less confirmed, in his mild-mannered, grumbling way, that he'd been concerned about the procedures that BP had put together for the temporary abandonment of Macondo. As the top Transocean person on the rig, it was his job to push back, and he did so, but on April 20 he had not behaved as a man with any sense of grave peril, and now, five weeks later, he told the investigators that he had no sense that disaster was moments away. So even as the story went out over the wires and across the country that this fellow on the Deepwater Horizon sensed that things weren't right, the investigators were also hearing that he felt sufficiently complacent that he was in the shower when the well exploded.

The Marine Board's investigation was going to be as tricky as drilling Macondo. The investigators were going to need to drill very deep. There would be trouble along the way. They could easily get stuck, or sidetracked. The truth was down there—somewhere.

High Pressure

Now came the top kill. It was, to use a favorite Thad Allen word, a "consequential" moment.

The operation would, if all went as planned, ram a load of heavy mud into Macondo's gullet, followed by a shot of cement. That would thoroughly plug the well. The relief would still come along and give Macondo another dose of cement down low, but that would be more like an insurance policy. Top kill was correctly named: This would terminate Macondo for all intents and purposes.

Or it could be a soul-crushing failure. There wasn't any middle ground. A mud shot into a dynamically flowing well either succeeds completely or fails completely. Failure at this point would feel like a disaster within the disaster.

Thad Allen worked around the clock to supervise the spill response while also dealing with the rituals demanded of a retiring Coast Guard commandant. In the span of a few days in late May, he oversaw a surge of Coast Guard personnel into the Gulf Coast, briefed his replacement as commandant, hosted 100 colleagues at an event at his home, presided over the retirement of the vice commandant, and finally, on May 25, relinquished his Coast Guard command. There was a picnic for 300; Allen's father, who had broken his hip in April, arrived from Arizona and attended the change of command in a scooter.

In a few weeks Allen would have to move out of the commandant's quarters, and he and Pam (still hoping to go to Ireland for two weeks in early August) had to pack up their belongings and put most of it into temporary storage. "I was walking around with an earpiece in, holding conference calls, telling the packers what to do," Allen said. The moment burned in his memory came one Sunday in mid-May at his grandson's seventh birthday party at a Chuck E. Cheese's: He had to go out into the parking lot to participate in a conference call, and the call lasted three hours. His family was not amused.

So it went for all the people involved, both in the government and at BP: The response was a twenty-four-hour job. Downtime was an illusion at best. Anyone who has been involved in a project of such intensity knows the feeling of never truly being off work. Normal domestic life becomes a ghostly backdrop to the vivid, hard, tangible reality of the project. In the military, there is a term: "battle rhythm." The people tasked with killing Macondo had been on battle rhythm for more than a month.

The top kill plan developed by BP would use the versatile, 312-foot Helix *Q4000* servicing rig, the vessel that had helped with the cofferdam, to pump mud into the well. A lot of hardware had to be in place. A dozen ROVs patrolled the deep, deployed equipment, and navigated around one another. This was more ROVs than had ever assembled in any given place on the planet. Dozens of vessels crowded the sea surface. Planes buzzed overhead. The jammed airspace posed severe hazards. Two days in a row, May 21 and May 22, aircraft nearly collided in the vicinity of the spill site. Two huge Transocean rigs, virtually identical to the Horizon, continued to drill relief wells just half a mile from each other. The Transocean drill ship *Discoverer Enterprise,* floating directly atop Macondo, collected a stream of oil from the well via BP's makeshift piece of hardware known as the Riser Insertion Tube Tool. A jumbo supply ship, the 381-foot *HOS Centerline,* stood ready to pump upwards of fifty barrels of mud a minute into the Macondo well. A second ship, the 300-foot *Blue Dolphin,* floated nearby as a mud-supply backup. All the vessels used satellite-based dynamic positioning. BP technicians monitored the vessels in the Simultaneous Operations (SimOps) section of the BP war room. Each vessel needed an escape route in case Macondo exploded again.

"As complicated as it looks on the surface," Kent Wells said in a BP video posted on the company website, "it's more complicated subsea."

The idea was to pump the mud down to the sea floor and into the blowout preventer through two different channels: the choke and kill lines. The three-inch lines enter opposite sides of the blowout preventer, one slightly higher than the other. The mud from the higher injection would create a kind of ceiling, or barrier, within the BOP. The mud from the lower injection, fired at a somewhat lower rate, would not be able to go upward because of this mud ceiling. So it would go down, theoretically—down into the casing, or into the annulus, or wherever the mud could go. Ideally, it would bully its way down to the reservoir. The pressure measured in the BOP should drop steadily, and if all went right, the accursed well would suffocate on mud. Then they could follow the mud with a massive dose of cement. And there would be no more oil spill.

There would be no more flowing Macondo well—just a hole that had been beautifully plugged.

That, at least, was the plan.

"We'll be able to pump much faster than the well can flow," Wells said. "It's about us outrunning the well."

The stakes were high, the situation fluid. There were a couple of new leaks in the riser. BP knew this, and the government knew this, but the news media and the American public never got the memo. The well's evolving leak was reported in an internal government update from the Unified Command: "New Development . . . The oil flow at the kink in the riser was initially leaking as two small plumes. Late last week, a third small plume developed, and on Tuesday of this week, a fourth small plume was observed, indicating that cracks are forming in the riser."

New leaks in the kink: That wouldn't have played well in the current media environment. The gloom had already thickened, congealed, turning into great, gloppy, emulsified streamers of pessimism and fear. A vast mousse of doom.

Every day the federal government tracked the public mood, not only via the mainstream media but also by dipping into the new "social media" and scrutinizing Facebook and Twitter comments. The government had set up a Facebook page for the Unified Command, as well as a Twitter account, a YouTube operation, a Flickr account, and the website Deepwaterhorizonresponse.com. But the public could never quite grasp what the Unified Command was, and how it was different (if it was different) from BP.

For example:

For Internal Use Only
Unified Area Command External Affairs Summary
Tuesday, May 25
Social Media Report
Statistics
Facebook followers: 23,798 (+494)
Twitter: Followers: 5,368 (+177); Lists: 370 (+10)
YouTube: Views 1,800,011 (+7,807); Videos 38; Subscribers 420 (+11)

Flickr: Photos: 433; Views: 80,903 (+5,406)
Website hits: 32,130,855 (+1,203,868)

Facebook Themes

—Lack of confidence that top kill will work.
—Animated discussions concerning some articles that were posted, claiming that explosions near the wellhead are causing the sea floor to collapse and have destroyed the BOP . . .
—Heightening impatience for the well to be capped.
—Continued concern that the information posted to our Facebook page is a BP PR maneuver.
—Continued passionate concern for the environment.
—Continued prevalence of concerns about a looming apocalypse . . .

Under "Trends," the Social Media Report noted "growing calls for the Federal government to intervene, without knowledge that the government is already providing an oversight role." The BOP-destroyed blog post cited in the Social Media Report was indeed rather apocalyptic:

AWFUL NEWS FROM THE GULF:
EXPLOSIONS COLLAPSE SEA FLOOR AT
DEEPWATER HORIZON WELLHEAD

A series of explosions appears to have collapsed the seafloor and blown up the BOP at the wellhead. Oil and gas are billowing out of a depression in the seafloor where the BOP used to be at an exponentially greater rate than anything seen before . . . the hole has been getting bigger and bigger and now it's like a volcano vent with the piping inside it probably a twisted mess.

The relief wells can't possibly stop this because anything they add will just be blown out of the volcano.

Now what?

The post ended with a caveat: "I'm seeking verification."

This was actually balanced and level-headed coverage compared to some of the stuff flying around the web. The plume became a brain worm in the head of hysterics. They knew that this was much, much worse than BP and the bootlicking government and the media enablers were revealing. The wellhead would crater; the seawater would descend into the seething cauldron below. Then, to hear one blog tell it:

> The water will be vaporized (turn to steam), creating an enormous amount of force, lifting the Gulf floor. It is difficult to know how much water will go down to the core and therefore not possible to calculate the rise of the floor. The tsunami wave this will create will be from 20 to 80 foot, possibly more. Then the floor will fall into the now vacant chamber. This is how nature will seal the hole.
>
> Depending on the height of the tsunami, the Ocean debris, oil, and existing structures that will be washed away on shore and inland will leave the area from 50 to 200 miles inland devoid of life. Even if the debris is cleaned up, the contaminants that will be in the ground and water supply will prohibit repopulation of these areas for an unknown number of years.

So maybe the sterilization of the Gulf Coast was an extreme example, but still, there were a lot of people who quite rationally wondered not only what would happen in the deep sea and along the beaches and in the marshes but also far inland if, and when, a hurricane grabbed the oil spill and rendered it airborne.

The week of the top kill was one charged with emotion. Many things were happening at once, and no one knew how anything was going to turn out. The top kill was scheduled to commence on Tuesday, May 25, the very day that a memorial service would be held in Jackson, Mississippi, for the eleven men killed on the Deepwater Horizon. Later in the week, Interior Secretary Ken Salazar would present a thirty-day report on the disaster that would impose a six-month moratorium on exploratory deepwater drilling, idling thirty-three drilling rigs and thousands of oil patch workers—a huge blow to the oil industry and to the Gulf Coast economy. President Obama scheduled a trip back to the gulf on Friday, at which time, if all went perfectly, the well would be killed, and perhaps he'd be able to stand on the beach in Grand Isle and talk about the cleanup and the way ahead, knowing that somewhere out there Macondo lay slain.

This was Issue One for the White House, something as big, as hot, as perplexing, as the war in Afghanistan, the long grind in Iraq, and the mopey economy. The administration had thrown everything at the spill: every agency, every scientist, every person who might possibly know how to plug a hole. But to the administration's torment, the public wanted Obama and his people to do more—somehow, someway. They did not grasp that the response was government-run, and no matter how many times the administration said, "We're in charge," the public never quite believed it.

A White House staffer asked the Interior Department to put together a list of things the government had done to help plug the well, saying that even a partial list "would be tremendously helpful in pushing back against the current press narrative." Suggested examples: Using gamma rays to peer into the blowout preventer, and arguing for a mud shot top kill rather than a junk shot.

In a later email, the White House staffer made an obvious point:

"Also understand we may not wish to claim credit for top kill approach until we see what happens."

The Obama administration had another fundamental problem in this particular crisis: Obama didn't seem mad enough. The American people wanted to see their leaders bring some heat, and they became impatient with the cool and collected Obama. The president still had that "No Drama Obama" thing going, that ability to ratchet crises down to the level of mere *challenges*. His equanimity was one of his great qualities, but suddenly he seemed like the guy who orders a fine Vouvray at a biker bar. People didn't want explanations and rationalizations from him; they wanted him to pop someone in the mouth. Make someone hurt. Instead he was aloof and professorial (*professorial* having become a pejorative in the age of right-wing talk radio and the Tea Party movement). He was disastrously unable to project a demeanor or temperament that could persuade the average American that he had his dander up. He didn't look like he had any dander at all. His press team tried to repackage him. They circulated an anecdote: In a meeting early in the crisis, Obama had grown so frustrated that he said to his aides, "Plug the damn hole!"

No one demanded more of the president than one of his political allies, Democratic strategist James Carville, famous as the "Ragin' Cajun." Carville went on CNN the night of May 26 and declared, "BP is not the equal of the United States government. This president needs to tell BP: 'I'm your daddy, I'm in charge. You're going to do what we say. You're a multinational company that is greedy, and you may be guilty of criminal activity.' It's time that we understand BP does not wish this thing well."

But Obama was more of an I'm-your-professor kind of guy. That very day, the president had been on the road, making a speech in California at the new plant of a company that builds solar panels. He noted that Steve Chu, "who, as you know, is a Nobel Prize–winning physicist," was on the scene in the gulf. Obama said nice things about solar energy and vowed that America would not fall behind China and Germany in the race to develop new energy technologies.

Good, green stuff—except when was he going to pop someone in the mouth?

Top kill had to work. Had to.

"We have the very best people in the world working on this," Doug Suttles told the news media. The COO tried to make clear to everyone that the top kill was no sure bet. But Tony Hayward went on TV and estimated that the operation had a 60 percent to 70 percent chance of being successful. Where he got his estimate was a mystery, but so was everything else about the subsea response. The news media reported what they were told (and at least they now had the plume to stare at), but everyone was scuttling for scraps of data the way the giant isopod searches the seafloor for rotting flesh.

The top kill would mean the termination of the Riser Insertion Tube Tool, the thing that looked like a dental instrument that had been sucking some of the oil out of the end of the riser. The device was a technological success worthy of a mild golf clap—*pock pock pock*—but no triumphal roar. BP believed that, at peak performance, it managed to capture 85 percent of the oil coming out the end of the riser, though the government found that dubious.

The flow rate had engineering implications. If it was much higher than BP suspected, the top kill probably wouldn't work. The mud pumps would never be able to outrun the well. In fact, the government had now come to realize that the 5,000-barrel figure wasn't even close to being accurate. Thad Allen, tired of the flow-rate controversy, had appointed the US Geological Survey boss, Marcia McNutt, still ensconced in Houston, to head a group of scientists to come up with a credible estimate for Macondo's emissions. This hastily appointed panel became known as the Flow Rate Technical Group. The group's initial calculations, based on just a few snippets of video provided by BP, showed that the well was leaking 12,000 to 25,000 barrels a day—*at least.*

The new estimate generated headlines and a great deal of confusion within the government and in the news media. Two teams within the group put the leak in the 12,000 to 19,000 range, and the government emphasized that number in its initial press release. But the third team, the "plume team," which focused on the plume video, came up with an estimate featuring a *lower bound* of 12,000 to 25,000 barrels a day, and no upper bound. So the range was really 12,000 to 25,000 or more. Government officials wrestled with how to present such confusing figures. This would not be the last time they faced the fundamental problem that different instruments and different techniques invariably produce different results even if they're looking at the same thing. Moreover, there are always error bars. Should the government be transparent about its uncertainties? Transparency dictated putting everything out there, uncertainties and all, but politics argued for the appearance of greater confidence.

At two o'clock eastern daylight time on May 26, the top kill began. It took awhile for the mud to reach the seafloor and enter the BOP. Almost immediately, the four little jets, or miniplumes, from the kink in the riser changed color: They had been black with streaks of white gas, but now they went mud-brown. The huge plume at the end of the riser, meanwhile, vanished from the ROV video feed amid flying mud.

The scientists and engineers were not simply watching the plume, however. They were studying pressure readings. And they knew exactly what to look for. The engineers had worked out the ideal scenario. Based on pressure readings taken at the BOP the day before the top kill, and based on what had been seen in these kinds of procedures—because wells have been killed with mud shots countless times, only never in such deep water—the observers hoped to see the pressure build briefly, and then drop steadily, eventually reaching zero as the mud rendered the well static.

And it started out just that way. Pressure up. Pressure down. Pressure dropping . . .

Alex Slocum, the MIT engineer, watching with other government scientists on the eighteenth floor of BP's Building 4, felt elated. *Fantastic,* he thought.

But then the pressure stopped dropping. It leveled off.

A series of emails lighting up in-boxes in the federal government tracked the pressure curve:

> 2:13: "So far it is a tie—the mud being pumped in seems to be mostly coming out of the riser. We have not yet hit a tipping point that indicates a killed well . . ."
>
> 2:24: "They seem to be having trouble outrunning the well (pumping in faster than well can spit it back out). They have increased the kill flow rate to 70 bpm [barrels per minute] . . ."
>
> 3:19: "The situation appears to be relatively steady. They have pumping mud at 60 to 70 bpm for the past hour and a half (about 4,500 [barrels] total). There are no indications that things are getting worse—but it is not clear that the kill has been (or will be) successful . . ."
>
> 4:35: "They just closed off all mud flow and are seeing whether the well will go stagnant. So far (4 minutes) the pressures are all going down slowly (a very good sign—but not conclusive)."
>
> 6:11: ". . . the pressures (kill/choke and below BOP) starting creeping upward . . ."

The jubilation leaked away early, followed by the confidence, followed by the guarded optimism. BP had stopped pumping mud, and before too long, the pressures had started to rise again.

Time to reload. They would try heavier mud. That evening Tony Hayward went on TV with a statement that was neither here nor there.

"The operation is proceeding as we planned it," he said.

The next day, with the status of the top kill ambiguous—the news media was quite thoroughly confused as to what was happening—Obama held a news conference in the East Room of the White House. It went poorly, the pundits declared. The president didn't have any good news to report. He was implicitly on the defensive from the moment he mounted the podium. The reporters were their usual, irritating, badgering selves. Obama assured the press and the public that the government was running the show, that it was directing BP, which was, nonetheless, the "responsible" party—an arrangement as befuddling as the idea that light is somehow both a particle and a wave.

It was hard to see what a winning move would be in the president's predicament. He had to convey command of the situation, but at the risk of claiming ownership of an epic fiasco. The rule still applied: No one looks good next to an oil gusher.

Almost every question was about the oil spill. (One question touched on Afghanistan, one on border security.) The spill, by this point, was threatening to consume all the oxygen in Washington. And Obama said as much:

"The gulf is going to be affected in a bad way. And so my job right now is just to make sure that everybody in the gulf understands this is what I wake up to in the morning and this is what I go to bed at night thinking about."

And then, with the press conference almost over, he added one more thought:

"And it's not just me, by the way. When I woke this morning and I'm shaving and Malia knocks on my bathroom door and she peeks in her head and she says, 'Did you plug the hole yet, Daddy?'"

Sound bite. Obviously that would be the screaming headline on the home page of the Drudge Report.

Did you plug the hole yet, Daddy?

What no one in the general public knew, throughout that Thursday of the top kill, was that the top kill had actually stopped—been put on hold—sometime Wednesday night. There was no mud going into the BOP anymore. Neither BP nor the government got around to mentioning this important factoid. The *New York Times,* which had a good source inside the black box, finally broke the story, saying that the operation hadn't worked as planned. Doug Suttles, appearing shortly thereafter in the daily news briefing, offered an equivocal analysis:

"Nothing has gone wrong or unanticipated," he said. "We did believe we did pump some mud down the well bore. We obviously pumped a lot of mud out the riser."

He added, "I believe this can work."

Finally, after a sixteen-hour gap, the mud kill attempt began again. This time, BP started with a junk shot, injecting golf balls and knots of rope and Happy Meal–like trinkets into the BOP to try to create some barriers against which the mud could push.

There was continued speculation in the media that the top kill was working. You couldn't see oil coming out of the well anymore. It was all mud, seemed like. Friday morning, Thad Allen made some comments to the effect that the operation had gained control of the pressure at the wellhead.

It was a rare misstep by the admiral: The media ran with it as good news, as evidence that the top kill was working. "Maybe I should have said that it appears at the current moment the pressure has suppressed the hydrocarbons," Allen told me later.

Obama visited Grand Isle, walked the beach, talked to the mayor and some other locals. A reporter shouted out, "Mr. President, how confident are you that it will be—that the leak will be plugged soon?"

The president answered, "All I can say is that we've got the best minds working on it, and we're going to keep on at it until we get it plugged."

But the best minds were going to have to keep working a lot longer, because the mud shots did not work—not the first day, not the second day, not the third day. "The plan is to continue until not even Count Dracula can survive," Chu said in an email late that Friday afternoon.

But Dracula survived. Finally, Saturday afternoon, BP and the government decided to halt the operation. Suttles announced the failure during the Unified Command media briefing, but in astronaut fashion did not dwell on the disappointment. He proceeded to mention the next operation so quickly and deftly that the unfocused listener might have concluded that the failure was actually a planned, incremental step toward a much better option involving the LMRP something or other:

"After three full days of attempting top kill, we have been unable to overcome the flow from the well, so we now believe it's time to move on to the next of our options, which is the LMRP lower marine riser package cap . . ."

Landry came up with an unnerving sound bite—"There is no silver bullet to stop this leak"—but otherwise remained remarkably upbeat given the grim news: "We also want to assure you we've had a very, very aggressive response posture," she said. "We're obviously right on the front lines in Louisiana fighting as the oil reaches the shore, but it is a tribute to everybody who has been working on this since Day One that we only have one hundred seven miles of shoreline oiled right now and approximately thirty acres of marsh."

BP came up with a theory: The well lacked integrity. Mud could have gone down the well bore and then escaped into the rock formation through those three rupture disks in the casing. No one ever thought, in designing the well with rupture disks, that one day these weak spots might make a top kill harder.

If the well lacked integrity, you could not kill this well from above. You had to contain the flow with caps and domes and contraptions until the relief wells hit Macondo sometime in August.

For now, it was the well that would not die.

Chapter 9

Repercussions

Now, the misery. The calamity had included, in spirit-sapping succession, a fatal blowout, a sunk rig, multiple leaks, a BOP intervention failure, a cofferdam failure, an unfolding environmental catastrophe, and now, after enormous hype, the top kill failure.

This was, for many people involved in the response, the worst moment of the entire disaster save the initial tragedy that killed eleven men. What made the top kill particularly shattering was that no one had made an obvious mistake. The engineers hadn't overlooked anything. They had storyboarded the operation to within an inch of its life. The government scientists had vetted the procedure. A few outsiders tried to build a case against Steve Chu, saying that he'd put limits on BP's ability to pump mud into the well, but that criticism never gained traction. The top kill had not been a wimpy procedure. The BP engineers had thrown 30,000 barrels of mud—more than a million gallons—at the well. There was nothing wrong with the mud, nothing wrong with the hoses or manifolds or pumps. The machinery on the boats had performed far beyond design capacity. No one blamed the golf balls or the Happy Meal stuff. Doug Suttles had been almost correct when he had told the news media during the top kill procedure that everything was going according to plan. The only thing that did not go according to plan was that it didn't work. This was one of those operation-a-success-but-patient-died deals.

There was no partial victory here. The top kill hadn't so much as bruised Macondo. The well was spewing oil as fast as ever, and BP began to suspect that the flow had gotten worse; that the mudding of the well had scoured it out, opened it up, like nasal mist up a nostril. This would be a debatable point. What's certain is that Macondo's hydrocarbons continued to shoot out the well as if they'd been fantasizing about being in open water for ten million years.

For everyone involved, this was rock bottom. No, worse: They were below the seafloor, buried in mud, like mollusks. The only virtue of the top kill was diagnostic, in the sense that a disease accurately diagnosed is an incremental step toward a cure. Now all those involved knew precisely how screwed they were. There was no magic pill. They would have to do the hard surgery of the relief well—unlikely to reach the target until late July or early August. This was but the end of May.

If you were someone like Marcia McNutt, working in that windowless room, you thought, *I'm never going home. I am a prisoner of this well.* Somewhere back in northern Virginia, she kept stabled a horse, Lulu, doomed to be riderless. She wanted to see her twin daughters ride in the California rodeo in six weeks, but she might never be allowed to see the sun again.

The hysterical hordes renewed their demands for the government to "take over" the oil spill, though it was never entirely clear what this would mean and how it would differ from the current arrangement as mandated by the National Contingency Plan. The Sunday before the top kill, Interior Secretary Ken Salazar had gone on NBC's *Meet the Press* and said that BP had busted every deadline so far. He then added, "If we find that they're not doing what they're supposed to be doing, we'll push them out of the way."

That baffled Thad Allen. The day after Salazar's remarks, Allen, facing reporters in the White House briefing room, said, "To push BP out of the way would raise the question, to replace them with what? . . . They just need to do their job."

There had never been a large reservoir of faith in BP as an engineering enterprise; that reservoir was now tapped out, a dry hole. Story lines in the national media rarely go into reverse; usually they harden. The bad oil company was now almost canonically evil and incompetent.

The comedy troupe Upright Citizens Brigade produced and posted on its website a brutal sketch titled "BP Coffee Spill" that went viral on YouTube, with more than eight million views by the end of June. We see hapless BP executives trying to deal with coffee that has spilled on a conference table and is threatening "the fish"—the sushi lunch. Increasingly daffy attempts to sop up the coffee fail spectacularly, and then more coffee is spilled, and everyone screams hysterically, and after forty-seven days the coffee is still everywhere.

South Park, the wicked, potty-mouthed TV cartoon on Comedy Central, captured the disaster perfectly, all the way down to the contagion of second-guessing. Soon after BP spills the oil, we see the arrival of Captain Hindsight, who can thrill onlookers with "the amazing power of extraordinary hindsight" ("They shouldn't have drilled in that deep of water, because now they can't get machines deep enough to fix the spill!"). He's useless, of course, in the fight against the spill, which gets inexorably worse: BP accidentally rips a hole into another dimension, demons and monsters pour forth, and a Godzilla-sized creature, "the dark and mighty Cthulhu," emerges, signaling three thousand years of darkness in which humans will "be driven to madness and made to service Cthulhu's cult as slaves." Tony Hayward, meanwhile, pops up repeatedly, saying "We're sorry" from a series of pleasant locales and vacation destinations, with a final apology coming as he warms his nude body on a bearskin rug before a crackling hearth.

The Obama administration needed to distance itself from BP, and not just with the random verbal slap. The administration realized that the public would never believe that the government ran this response. It needed to find a way to pop BP in the mouth *now*.

On Monday, two days after the top kill failure, the administration put out word that Attorney General Eric H. Holder Jr. would go to the Gulf Coast to confer with federal and state prosecutors. He wasn't going to the gulf to soak in the healing waters. The Justice Department had already told BP to save all its internal correspondence and paperwork, pending a possible lawsuit or even criminal charges. BP obliged, as company officials hoped to avoid anything so disruptive as search warrants and computer seizures.

Next, the administration put the kibosh on the Landry-Suttles news briefings. There would be no more joint briefings from the government and BP. In effect, the White House fired BP from the communications team. The idea was to obliterate the image of the government and BP working together as equals. This could no longer look like a "partnership" or "collaboration." The government had to show that these were not equal institutions; that the government was the alpha to BP's beta.

Rear Admiral Landry largely disappeared from the story, as she rotated back to her job running the Coast Guard's Eighth District. Coast Guard Rear Admiral James Watson replaced her as the federal on-scene coordinator for the oil spill. More significantly, from now on, Admiral Thad Allen would deliver the daily briefing by himself, no matter where he was, whether that be Washington or New Orleans or Houston or aboard a drilling rig in the Gulf of Mexico. (I later learned that he sometimes held the briefing on his cell phone as he sat at the kitchen table of his temporary residence in Fairfax, Virginia.)

The White House explained that comments by BP had contributed to the decision to end the joint news conferences. The administration had lost confidence in the candor of the company. The White House offered one example: The next subsea maneuver called for the shearing of the riser just above the blowout preventer. The administration said that could increase the flow by as much as 20 percent in the short run, until a new containment cap could be put in place. BP had said the increase might be more modest, just a few percentage points. The White House didn't want to be contradicted by BP anymore.

"We've been increasingly frustrated with BP on matters of transparency," an administration official told me. "We're not going to stand there while BP says there's not going to be any increase in flow rate when they cut the riser." In other words, we can no longer stand next to—or *stand*—this dishonorable oil company.

The move to end the joint briefings pleased BP's critics. Ed Markey, the Democratic congressman with a gift for the sound bite, said one day, "BP now stands for *Bad Partner*. If it's BP's spill, it's *America's ocean*. BP is interested in its *liability*, and the US government is interested in the *liveability* of the gulf." (A sound bite trifecta!) Markey cut to the main point: "I think that BP has not demonstrated a level of competence or trustworthiness that merits having the US government standing next to it at press conferences."

Obama, meanwhile, experimented with a butt-kicker persona. Appearing on the *Today* show, he told host Matt Lauer, "And I don't sit around just talking to experts because this is a college seminar; we talk to these folks because they potentially have the best answers, so I know whose ass to kick."

It didn't sound quite right, this note of profanity, coming from the mouth of the Great Equilibrator. The president was like a middle-aged man trying on a leather jacket, knowing full well he'd feel more comfortable in tweed.

The administration, not finished with its abjuration of BP, announced that it would make a basic change in the Unified Command's website, deepwaterhorizonresponse.com, the clearinghouse for information about the spill. BP reimbursed the government for the website and collaborated in its content development. The administration decided it was time to switch to a government-only website, one with a dot-gov address instead of a dot-com address.

This change in optics (to use a favorite Beltway-insider term) had elements of an optical illusion. The US government was not merely standing next to BP at news conferences and collaborating with BP on a website. The government was working with BP every day at multiple command centers under the rubric of the Unified Command. Government officials and scientists remained embedded at BP's headquarters in Houston, with more officials arriving, more scientists getting involved, and the relationship intensifying by the day. The Coast Guard, the MMS and the US Geological Survey had people in the hallway on the third floor right outside the BP war room. They sat in meetings together, talked to one another in conference calls, worked together on plans. BP executives and top administration officials signaled to one another by email what their communications strategy would be at any given moment.

Thad Allen recalls this delicate period: "What you had was a simultaneous political response and an operational response to this thing. They're not going to be always congruent, but you handle both of them. That was my job. The White House wanted to be seen as strong. This was right after the 'kick the ass' and 'boot on the neck' thing."

The response to the spill had multiple fronts, and the government could tighten its supervision of shoreline defense and the vast fleet of boats skimming oil on the gulf surface. Some 45,000 people and nearly 7,000 boats would eventually be involved in this unprecedented operation. But the plugging of the well—the subsea response—posed a special challenge for the government.

After top kill, Steve Chu dove into the problem with renewed vigor and intensity. He got down in the weeds of this Macondo well. He studied the diagrams of the well, did calculations on the design limits of the hardware, tried to deduce the flow rate from pressure readings. He would come to know this well to the point that he could find his car keys down there.

Most of the government scientists in Houston worked out of the eighteenth floor, fifteen floors above the BP war room. They could go downstairs to the war room proper but mostly kept to their fiefdom high in the building.

Chu's relationship with BP was cordial but not warm. He grew impatient with the pace of information coming from BP. He was astonished by the lack of diagnostic equipment deployed by BP and its contractors in the deepwater environment. The blowout preventer, for example, cost many millions of dollars and yet somehow came equipped with only a single pressure gauge, near the base.

"It was crazy. There were no indicators that you could use to tell that all those multiple valves were closed and locked," Chu said. "We had to strongly encourage them in many instances to stop and take measurements whenever possible, to figure out how to get instruments down there. It was not in their DNA that that was the way you do things."

Chu's emails to colleagues were not terse like most of Thad Allen's emails (one would think the admiral was concerned that he was being charged by the character), but rather were lush with scientific insights and equations. For example, one day in June, he was sharing his thoughts on how to discern the flow rate from pressure readings on the new cap on the well:

> My understanding of the basic physics is a steady state analysis where the change in d(kinetic energy density)/dh distance is determined by a buoyancy factor (the *g term), the pressure difference dP/dh, the friction terms due to the vent structure including the inlet and exit structures and the 3" valves, and some "equivalent friction" term due to the change in density as the oil/gas mixture meets the seawater. Details such as two component and single component flows up to sea level, the temperature profile up the 5,000 ft riser, etc. did not matter that much. Also, we would like to get a better feel for the sensitivity of any assumed geometry at the

> bottom of the Top Hat skirt. With the previous assumption of an asymmetric gap, there was only a ~25% effect (assuming three of the vents were open and 4" ball valves) of the amount coming out of the Top Hat and no correction to the oil/gas seawater interface issue. There is still an unresolved issue of how to deal with the interface going from the oil/gas mixture in the Top Hat to seawater.

And so on. Hairy stuff, totally incomprehensible to the average human being. As Chu jotted complex equations on a whiteboard, his fellow government workers were amazed at the depth of his involvement and began to wonder, *Who's running the Department of Energy?*

One day in June, Chu sent his energy subordinates a playful email with the subject line "You're in it now, up to your neck!" He explained that the new relationship between the government and BP put more responsibility on the Energy Department scientists to vet the subsea operations:

> There is a scene in the *Guns of Navarone* where Gregory Peck (Captain Mallory) was talking to David Niven (Corporal Miller) just after Irene Papas (Maria) shoots the traitor Anna. ". . . your bystanding days are over! You're in it now, up to your neck! They told me that you're a genius with explosives. Start proving it!" Probably no shaped charges to be used on this mission, but the rest rings true . . . Steve.

The administration had scrambled all of its jets, had sent Cabinet secretaries into overdrive, had written a blank check to the agencies to do whatever it takes to deal with the spill. And yet on the ground, on the beach, on the sea, at the bottom of the gulf, the results remained miserable to behold. The oil was really coming ashore now, with confirmed impacts on the Chandeleur Islands, Whiskey Island, Trinity Island, Raccoon Island, South Pass, Pass a Loutre, Fourchon Beach, Grand Isle, Elmer's Island, Brush Island, Pilot Bay, Timbalier Bay . . .

An instantly iconic image would soon circulate on television, the Internet and the front pages of newspapers: a small bird, its species unidentifiable, so completely covered in thick brown oil that it looked as though it were made of milk chocolate.

An ABC News/*Washington Post* poll released June 7 reported that 69 percent of the public rated the federal response to the oil spill negatively—higher, even, than the 62 percent who rated the federal response negatively two weeks after Hurricane Katrina. The pundit class continued to demand more emotion from Obama. Reporters asked the press secretary, Robert Gibbs, for specific evidence that Obama was hot under the collar:

> Q: You said earlier that the president is enraged. Is he enraged at BP specifically?
>
> Gibbs: I think he's enraged at the time that it's taken, yes. I think he's been enraged over the course of this, as I've discussed . . .
>
> Q: Frustration and rage are very different emotions, though. I haven't—have we really seen rage from the president on this? I think most people would say no.
>
> Gibbs: I've seen rage from him, Chip. I have.
>
> Q: Can you describe it? Does he yell and scream? What does he do?
>
> Gibbs: He said—he has been in a whole bunch of different meetings—clenched jaw—even in the midst of these briefings, saying everything has to be done. I think this was an anecdote shared last week, to plug the damn hole.

The government had to do something more to stop this. So it was that the Jindal-Nungesser sand berms got a new lease on life. The sand berms had effectively been killed by Thad Allen and the various government agencies that had reviewed the concept. Allen had approved only a "prototype" berm, hoping to throw a sop to the Louisiana contingent and get the issue to go away. It would not be a cheap job, costing $16 million (BP was asked to pick up the tab), but it was nothing like the massive project for which the Louisianans had hoped.

Then Obama threw the more ambitious plan a lifeline. On May 28, during the president's second visit to Louisiana, he met with local and state officials in a small conference room at the Coast Guard station on Grand Isle. On hand were Jindal, Nungesser, New Orleans mayor Mitch Landrieu, Alabama governor Bob Riley, and Florida governor Charlie Crist, among others. Nungesser gave the president an earful. So did Jindal. They all wanted more action, faster. Should have done it three weeks ago! Don't just sit there . . .

It's very hard, if not impossible, to say no to extremely vocal, red-faced elected officials who are sitting there looking at you like you're some University of Chicago constitutional law professor with no freakin' idea what it's like to have oil on your hands and on your boots and in your etouffee. Obama, while not committing to the berms idea, turned to Thad Allen and asked him to . . . you know . . . take a look at this berms idea. Allen said, Well, that's tricky, can't necessarily do that in a snap. (It was the Friday before Memorial Day weekend.) Obama asked if he could get it done in the next week. It's not clear whether Obama understood that the berms had already been vetted; that this had been discussed thoroughly by relevant government agencies and found unhelpful at best. The president had put the admiral on the spot—"hamstrung" him, as Allen later told the presidential commission.

The admiral followed through and did what he was told, and the following Wednesday a bunch of experts stood on a stage and hemmed and hawed about the sand berms while Nungesser and Jindal and other Louisiana officials stared at them from a few yards away. There was mild support, at best, for the berms, but no one said the berms were the dumbest idea since the hairstyle known as the mullet, and thus when it was over, Allen went with the flow—and approved the plan. BP would pay for it; the dredges would get to work.

The project would ultimately cost $220 million, according to the presidential commission staff, and trap "not much more than one thousand barrels of oil." The berms, the staff concluded, were "underwhelmingly effective, overwhelmingly expensive." But the staff also recognized the political environment that the administration faced.

"As a result of the surrounding context, the bar for approval of the project was set low, and the State of Louisiana cleared it," the staff report stated.

Put a Lid on It

In trying to kill Macondo, the engineers had already gone through plan A and plan B and plan C and plan D—the BOP intervention, the cofferdam, the Riser Insertion Tube Tool, and then the top kill—and were now simultaneously working on plans E, F, G, and so on. There were so many parallel plans that they might soon need to switch to the Cyrillic alphabet.

If Macondo could not be killed just yet, it could still be contained. Marcia McNutt predicted that if they could contain all the spilling oil, the story would fall off the front page. Then the ultimate killing of the well by the relief well would be on page A10 or whatever. "A footnote," as she put it.

Containment was BP engineer Richard Lynch's job. The failure of the top kill was his cue. Lynch's sprawling corps of engineers had been staging equipment, storyboarding procedures, and going through "finite element analysis" to ascertain potential failure points. His teams had engaged in "destructive engineering" to see what might break or clog or tangle. After the top kill, Lynch said, "We were geared up. We had everything on the flight deck, we were ready to take off."

The gear included a new cap for the well, an inverted funnel known as the top hat. The first order of business was to cut the riser. Recall that, after the top kill, there were still two leak sources hundreds of feet apart. One leak was from the end of the riser: That was the plume. The other leak was a cluster of four small plumes along a crack in the riser kink a few feet from the top of the BOP. The engineers needed to transform two leaks into one leak. They would need some very large scissors for this. There was just such a tool at a yard in Port Fourchon: a pipe cutting shear that the technicians called BAS, for Big Ass Shear. It was normally used for the demolition of rigs in shallow water. It looked like something the comic-book artist Jack Kirby would have dreamed up; hardware fit for the Fantastic Four. On Tuesday, June 1, the ROVs at the sea floor used this frightening contraption to chomp through a section of the riser a short distance from the top of the BOP stack.

The next day, the ROVs attempted a more delicate operation. A submersible used a diamond saw to slice into the riser just above the blowout preventer. The saw became stuck. "Anybody that's ever used a saw knows every once in a while it will bind up," Thad Allen told the media. The engineers finally decided to bring back the Godzilla shears. They chomped the riser, but the shears left a jagged chimney, nothing like the clean cut they'd wanted for the new cap. It would have to do. The engineers had anticipated that they might not get a good cut, and so they had another hat, top hat no. 4, standing by for just such a circumstance.

Now Macondo had a single leak, gushing profusely, dramatically, straight up from the BOP and out the jagged chimney. The ROVs took high-definition video. (The government scientists would conclude that the flow had increased only a few percentage points with the removal of the kinked riser—pretty much exactly what BP had said before getting the boot from the joint news conferences.)

On Thursday, June 3, the next big operation began. Anyone in the world with a computer and a decent Internet connection could watch—via BP's live feed from twelve ROV cameras—the underwater dance in which top hat no. 4 found its perch on the Macondo chimney. Technicians lowered the hat next to the gusher, and then eased it laterally over the plume, the way the cofferdam should have been deployed.

The new cap found its seat on the plume. The oil went up a riser to the drill ship *Discoverer Enterprise,* whose intrepid crew members would have to live and work directly atop the dangerous well. The ship could process about 15,000 barrels of oil a day while flaring the separated gas. The oil would be unloaded to a tanker and shipped to shore.

The new containment system worked splendidly from an engineering standpoint—the *Enterprise* ramped up quickly to 15,000 barrels a day of oil captured from the well—but there was a problem: The plume hadn't gone away. The top hat didn't capture all the oil, not by a long shot. The plume, though partially diverted to the surface, was still very much there, only in changed form. It could be seen by the whole planet, twenty-four hours a day, via the ROV feeds: a furious, frothing, riotous cloud of oil and gas exploding from underneath the loose-fitting cap and from vents at the top.

This transmogrified plume continued to camp out on the cable news and the Internet sites. For BP critics, this did not look like "success." In fact, it seemed to shout "They still didn't fix the [expletive] problem!" It was as though the scale of the operation was wrong, kind of like in *This*

Is Spinal Tap when, during the heavy metal band's big onstage production number, a Stonehenge monolith descends, and it's only a couple of feet tall. There was more oil than hardware, more gusher than containment capacity. Top hat no. 4 was not equal to Macondo. For the engineers who designed it and the technicians who deployed it, the top hat was a triumph of ingenuity and resourcefulness in an extraordinarily challenging environment; for the people who just wanted the plume to go away forever, the top hat looked kind of wacky, kind of jury-rigged, what with the way it was tilted off the vertical, wobbling slightly, and almost vanishing amid the mad clouds of oil and gas. The plume looked like it would soon dissolve the top hat entirely, the way Coca-Cola can dissolve a human tooth.

There was just too much oil. The flow was overwhelming. But what was the flow rate? The Flow Rate Technical Group had been another ad hoc addendum to the spill response, and now it assumed greater importance. The group spent weeks in late May and throughout the first half of June trying feverishly to come up with a solid estimate of the Macondo flow. The academics on the team—scattered around the country—had been subsisting on the most pitiful video scraps from BP. They were up against the clock, doing accelerated science of an inherently imprecise nature. At one point, NOAA scientist Bill Lehr, angry about a glitch in an online press release, obliquely threatened to quit. One scientist on the team protested that another had been making outrageous statements in the press: "He seems to have lost his common sense."

And the numbers kept changing.

Just four days after the government released the estimate of 12,000 to 25,000 barrels, new video, better video, of the leak from the end of the riser—taken prior to the riser cut and the new top hat—gave the scientists a chance to revise the estimate. By June 10, the government had put out a new estimate of 20,000 to 40,000 barrels a day, with a best estimate of 25,000 to 30,000 barrels. If you were a reporter, you had a lot of numbers to play with. During news briefings, there was some talk of 50,000 barrels a day, possibly. The numbers were all over the place. This was like trying to spear a martini olive with a tiny plastic sword.

"The administration is coming under increasingly heavy pressure to come up with a sound number of the flow rate," Chu told his colleagues by email on June 13. He added: "Science is not a democracy, and we hope to drive toward a single estimate."

But here was a problem. As Chu noted, science isn't a democracy, at least in the sense of operating by a show of hands—but it does have a democratic aspect to it. The truth cannot be established by fiat. It cannot be dictated from on high. The "argument from authority" carries no weight in scientific circles. Having tenure at an elite institution does not make one's pronouncements necessarily more valid than the conclusions of a lowly grad student. Settling on a single flow rate number would not turn preliminary estimates into hard facts. If the scientists decided to go with a single number, they could set themselves up for an embarrassment down the road if new information changed the picture.

BP had turned over some high-definition video, so vivid the observer could nearly taste the oil as it spurted from the well. What a fire hose this Macondo well was. The plume people ran the numbers again, used their techniques for tracing a single particle in a plume and factoring in the diameter of the aperture and the ratio of oil to gas and so on. Meanwhile, the Woods Hole scientists, who had been brushed back in early May, finally got a chance to sail to the source and send down some acoustic instruments that obtained a flow rate estimate. Their conclusion: this was 59,000 barrels a day.

The X-Men and the other government scientists worked on their own estimate, wholly independently of the Flow Rate Technical Group. Chu's people insisted that BP insert a pressure gauge into the top hat and monitor it with an ROV. The ROV stared at the pressure gauge; government officials monitored the readings. Then lightning struck, and not metaphorically. An actual bolt of lightning hit the derrick of the *Discoverer Enterprise*. The derrick caught on fire. The crew put it out quickly, but in the meantime had to stop siphoning hydrocarbons. The sudden termination of the operation meant that no longer were 15,000 barrels of oil going up to the *Enterprise.* The pressure in the top hat should have spiked. But it had barely changed. The obvious conclusion was that the pressure gauge was never working to begin with.

If you were Thad Allen, you started to get tired of all the numbers flying around, the multiple teams with their divergent estimates, the science-by-media-interview, and the estimates constantly ratcheting upward in a manner that seemed to indicate that the government could never quite make up its mind. With the scientists preparing yet another estimate, Allen wrote his colleagues:

"I'm having a disconnect . . . We have agreed to produce one number for the federal government. Release of serial information creates confusion and erodes credibility . . . What are we trying to achieve with this release? Correcting misinterpretation of announced ranges?"

Finally, after all the scientists on all the different teams threw their best numbers into the mix, the government produced a headline-grabbing estimate: 35,000 to 60,000 barrels a day. This was still a wide range, hardly meeting anyone's goal for a single, firm estimate. But the upper bound caught everyone's eye. It was precisely the flow rate that BP, back in early May, had said would be the "worst case"—and implausible—scenario for a completely uncapped well.

The worst case was now staring everyone in the face.

Richard Lynch's operation at BP ramped up containment capacity as fast as it could. With the top hat working, his team came up with a second containment operation that would pull oil and gas out of the well via a line emerging from the old blowout preventer. (Richard Garwin had proposed this early on, but Lynch says this is not a novel idea and had been recognized by BP as an option from day one). The servicing rig *Q4000* handled that operation, and it went quite well, siphoning roughly 8,000 barrels of oil a day from Macondo. The *Q4000* couldn't store any of the hydrocarbons, but it could flare them, both oil and gas simultaneously.

That meant, between the top hat and the new *Q4000* operation, upward of 23,000 barrels of oil a day were being contained rather than polluting the gulf. On some days, the number was even higher: more than 25,000 barrels. Incredibly, though, the plume still didn't go away.

Steve Chu came up with a simple way of deducing that the flow had to be much higher than BP had ever suspected. He looked at two videos side by side as they played on laptop computers: One showed the plume gushing around the top hat prior to the *Enterprise*'s coming online to collect any oil; the second showed the plume after the *Enterprise* began gathering 15,000 barrels a day. Chu could not tell the difference. The plume looked identical even with 15,000 barrels of flow subtracted. He explained the concept in an email to a colleague:

> Recall my little experiment I did while in Houston? By staring at two simultaneous videos of what was coming out of the Top hat at a collection of 15,000 bpd and 0 bpd, we could not tell the difference.
>
> Define the oil coming out of the well as R. While collecting 15,000 bpd, we have 15,000 + fR = R, where f is the fraction coming out of the top hat during production of 15,000 bpd. Suppose we couldn't tell by eye that the flow has decreased by less than 1/3. Then, it is possible that the amount coming out of the top hat was 2/3 R, and that tells us the lower limit of the oil flow must be 45,000 bpd. In order to get an estimate of 30,000 bpd, we would have to conclude that it is impossible to tell by eye that the flow has decreased by a factor of 2. Nobody, not even the plume group, would make that claim. This observation, which Kent Wells argued was "not data," gives a reasonable lower limit to the total flow.
>
> While collection 22,000 bpd, people say they see a noticeable difference. Using the same logic, we can get a rough upper bound.
>
> 22,000 + fR = R. In this case, suppose the rate out the top hat has decreased by a factor of 2. Then R ~ 44,000 barrels. If, on the other hand, we are sensitive to a decrease to 2/3 R, then R = 66,000 barrels.

On June 7 Obama held a Cabinet meeting and voiced a desire to provide BP with "in-depth guidance that reflects our concerns with the capacity and redundancy of the spill containment plan," as Allen put it in an email to Coast Guard colleagues. The last line of the email is instructive: "This directive to BP is also going to be the basis for an external communications strategy." Meaning: We are going to tell BP to expand its operation further, add capacity, add redundancy, and as we do this, we're going to tell the news media and the public that we're doing it. And the world will see that we're kickin' some ass!

A June 8 letter from Coast Guard Rear Admiral James Watson to BP's Doug Suttles explicitly stated that BP needed to get cracking on containment:

"Now that the so-called 'top hat' containment system has begun to capture and recover some of the oil escaping from the wellhead, it is imperative that you put equipment, systems, and processes in place to ensure that the remaining oil and gas flowing can be recovered . . ."

Suttles responded the next day, laying out the containment plan put together by Lynch and the others. Although the letter was hardly argumentative, it did note, gently, that elements of the plan had been in process for "several weeks" (meaning, we don't really need you to tell us to do what we already have been doing). The company now had a blueprint that called for four more drill ships and tankers joining the fleet at the spill site. BP would build two permanent risers that would increase collection capacity. The new plan should allow it to capture 40,000 to 50,000 barrels a day, Suttles wrote.

That was a lot of containment capacity, but it still wasn't enough, the government declared. What would happen in a tropical storm? Hurricane season had arrived, and the war room engineers needed to figure out what to do when the big blow came in. The *Enterprise* and the *Q4000* would have to skedaddle in a hurricane, disconnecting from the well and letting the oil gush freely once again into the gulf. The industry orthodoxy regarding hurricanes is simple: You must evacuate. A big rig can ride out winds of fifty to sixty miles per hour, but gulf hurricanes can easily double that. A storm can strengthen from a tropical depression to a killer hurricane in just a couple of days. The ships typically sail to calmer waters many days in advance of a storm. The *Enterprise* and *Q4000,* moreover, couldn't simply pack up one day and sail out of the way of a storm. Pulling up the riser to the *Enterprise* was a three-day operation. A hurricane would likely result in about ten days of no containment.

The BP plan had too little redundancy. Watson wrote back to Suttles with the message that the company had to do better and devise a plan that could handle 80,000 barrels of oil a day.

Suttles wrote back to Watson: We'll do it. We're bringing in two-thirds of a mile of specialized hose from Brazil, we're going to do repairs on key parts of the blowout preventer, and so on. By mid-July, he promised, BP should be able to capture up to 80,000 barrels. But Suttles added a warning: This is dangerous stuff, and we can't promise to capture every drop of oil coming out of the well. We're not going to get anyone killed out there.

"Several hundred people are working in a confined space with live hydrocarbons on up to 4 vessels," Suttles wrote. "This is significantly beyond both BP and industry practice."

The White House could now go to the public and say that it was forcing BP to do what the oil giant really would rather not do. Robert Gibbs said this explicitly in his press briefings. On CNN, the press secretary emphasized, "They're drilling not only a relief well to put a permanent end to this crisis in the gulf but they're drilling a second relief well at the cost of one hundred million dollars, not because they wanted to but because the government directed them to. They're increasing their containment strategy on the surface so that we can pump more oil up, not because they wanted to but because we directed them to."

Obama to BP: I'm your daddy.

The administration apparently wanted the public to perceive the two parties as fundamentally adversarial, with the government bending BP to its will at every turn. The situation on the ground wasn't quite so dramatic. BP would propose, the government would review. The government would push, BP would respond. The company, oddly enough, conspired in the illusion of government as master. Both sides got something out of this "external communications strategy." The administration could boast that it had full command of the situation and was making BP do its bidding, and BP could point to the government and say, Look, this is government approved, they're running the show, we're just law-abiding citizens doing our civic duty, and so on. The arrangement gave BP a measure of political cover from some of the second-guessing and the accusations of ineptitude and deviousness. It might also come in handy in a courtroom if and when the government sued BP for the amount of oil spilled into the gulf.

As a government official informed me by email, "The 'orders' that the gov't was giving BP were for the most part what BP had planned to do anyway. The vast majority of what Gibbs called 'orders' to BP were the gov't team in Houston and the BP engineers spending days in reviews coming to a very amicable agreement on the path forward, communicating that to the higher up chain of command (Chu, Allen, etc.), and then what we had agreed coming back to BP as an 'order.'"

Thad Allen said that by this time in the crisis, BP was long past the point of resistance whenever the government made a demand. "At that point, they were beyond arguing with us," he said.

It's impossible to know if BP would have eventually built all that containment capacity without the timely government jabs to the ribs. BP was going bigger, but it might not have gone big enough fast enough without the government's new flow rate estimates. The company never thought that Macondo could produce so much oil. The veteran engineers running the response were sure that the flow rate was a small number. They thought the gamma ray people were losing credibility with these outlandish new flow rate numbers. But the continued investigation of the flow rate seemed to support the worst-case scenario. Every new estimate reached the same conclusion: Macondo was a monster blowout.

The decision to go big on containment required a change in hardware at the top of the well. To capture upward of 50,000 barrels of oil, BP needed more avenues by which the hydrocarbons could be pulled. It needed a new cap on the well. The top hat would have to go. Lynch's group drew up a plan that, if implemented (and this was a big if), would replace the top hat with a new, tightly fitting cap, one that could seal the well and divert the oil via two new lines to surface ships.

This cap was called the "sealing cap," or, for those wanting to be precise, the "3-ram capping stack." This new stack of hardware would be—to use, once again, that favorite Thad Allen word—consequential.

Chapter 10

A Disaster Site

Just as BP was finally getting its footing on the seafloor, and just as the government scientists were feeling more confident and assertive in their role as BP's ad hoc board of technical supervisors (BP surely had less flattering descriptions), events far from the Gulf of Mexico threatened to undermine the subsea response.

Many of the people who mine money for a living, who run the profit extraction industries, who can pump billions of dollars in capital through digital pipelines and then choke those lines at will, had decided that BP had no future. BP stock that had been trading at $60 a share on April 20 had fallen below $40 a share after the top kill failure and showed no sign of leveling out as it headed toward $30. Wall Street increasingly looked at BP as if it were something the cat had left on the front stoop. This was an existential crisis for BP; a rumor of bankruptcy was now in the wind. Financial institutions that once counted BP as a favorite client now shunned the company. As the days ticked by, the prospects for BP looked bleak, and the other oil giants began circling, wondering if BP might be vulnerable to a takeover. They were ready to dismantle BP if necessary. The mighty oil company could soon be little more than scrap metal.

BP had a technological problem and a political problem. It could neither plug the well nor stem the flow of harsh rhetoric coming out of Washington. Some of this was just chatter that could be ignored, but then on June 9, Interior Secretary Salazar said in a Senate hearing that BP should be forced to pay the lost wages of oil rig workers—including the employees of other companies—sidelined by the government's moratorium on deepwater drilling. His logic: The rig workers were out of a job because BP had screwed up colossally. After word of Salazar's comments spread, BP's stock fell off a cliff: a 16 percent drop in a single day.

Salazar's comments received a White House endorsement. Press Secretary Robert Gibbs, asked about the statement, said, "The moratorium is a result of the accident that BP caused . . . Those are claims that BP should pay."

BP put out a statement on June 10: "BP notes the fall in its share price in US trading last night. The company is not aware of any reason which justifies this share price movement." That same day, the credit default swaps on BP—essentially, financial bets that the company would go bankrupt—shot up in price. Within two days, they reached escape velocity. The credit default swaps went illiquid, meaning that no one on Wall Street was willing to take the other side of the bet *at any price*. Wall Street had made its prediction: BP was going under.

Until this point, BP had generally avoided getting in a verbal tit for tat with the US government. Indeed, the company had generally responded to verbal attacks from government officials with silence, or muted, bland comments. But the liability issue was a deal breaker. From BP's perspective, the US government had idled the deepwater drilling workers. The US government had imposed the moratorium on drilling and could lift it at a moment's notice. BP heard in Salazar's comment the possibility of infinite liability. This was going to cost the company something on the order of $30 billion, or more—a staggering sum that few corporate entities in the world could possibly afford. Whatever it cost, BP needed to have some sense that this was a finite disaster. It needed this disaster capped, sealed, plugged. Somehow.

"At some point you have to draw a line," a company insider told me. "BP will not hand over a blank check to anyone, whether it's the administration or an independent mediator."

Not everyone, as it happened, hated the company formerly known as British Petroleum. The British, for example, were rather fond of it. The government pension system was heavily invested in BP stock, and with that stock tanking, this was having a potentially devastating impact on British retirees. A number of British politicians cried foul, including London mayor Boris Johnson, who criticized "anti-British rhetoric," "buck-passing," and "name-calling," and Lord Tebbit, a former

Conservative Party leader, who said that Obama's actions during the spill had been "despicable."

The newly elected prime minister, David Cameron, hectored by his constituents for not sticking up for BP, spoke to Obama about British concerns and later suggested publicly that perhaps the Americans should ease up a little.

The administration needed to sort this out. Obama summoned BP for a meeting at the White House—a command cloaked as an invitation. The optics had to be handled carefully. Obama had never met with Tony Hayward or even talked to him. Their lack of communication had become an issue in the media. For nearly two months, both men had been up to their eyebrows in spilled oil, but the Obama team had never felt it appropriate to put the president in any kind of proximity to Hayward, or even link them up by phone.

In another, kinder universe, Hayward might have been Obama's type. He was, after all, a credentialed geologist, something of a boy wonder, someone who had worked his way up in the company not through political connections but through force of intellect. He was, like Obama, a creature of the meritocracy. But Obama thought before he spoke, a talent Hayward never mastered. Hayward demonstrated during the oil spill crisis an astounding gift of gaffe. He was so accomplished at uttering inappropriate remarks at inappropriate moments that one would think he practiced in front of a mirror.

The classic Haywardism, immaculately timed, had emerged from his lips on May 30, at the nadir of the company's technological fortunes, when the top kill had just failed. Speaking casually into a camera, Hayward said: "There's no one who wants this thing over more than I do. You know, I'd like my life back."

No one paid attention to what he'd said to set up that comment: a statement of empathy for the people of the gulf and the disruption the spill had caused. Life is cruel, and the mediasphere crueler yet, and thus everyone focused on the core statement, the five-word nugget of self-pity: *I'd like my life back.*

Hayward, against all common sense, seemed to believe he could turn the tide of public opinion; that by force of personality, he could win people over. He did not understand the essential dynamic here: He was going to be the bad guy in this story. He thought he could assuage the nation's pain with the right words, but, sadly, words were things he was—you know—not so good at. He was a geologist, not a public relations wizard. At times, the poor man seemed to have rocks in his head, as when, more than three weeks after the blowout, he downplayed the disaster in an interview with the *Guardian* newspaper. "The Gulf of Mexico is a very big ocean," he said. "The amount of volume of oil and dispersant we are putting into it is tiny in relation to the total water volume." It was classic Hayward timing, coming just after the world had gotten its first look at video of the Macondo plume.

Hayward had another chronic problem as he toured the gulf, talked to cleanup workers, met with local officials, and chatted with the news media: He was conspicuously English. In a crisis, people can get tribal. Americans do not cotton to anyone who puts on airs, and Hayward, with his posh accent, sounded like someone who might tell you that you were using the wrong fork. He didn't look like someone who'd ever gotten oil on his hands. Did he know how to change a lightbulb, much less a blowout preventer?

The simple fact is that Americans pretty much hated him the same way they hated the oil spill itself. Obama could have more easily met with the crackpot running North Korea than with Tony Hayward. If someone snapped a photo of Obama shaking Hayward's hand, it would have looked as hinky as the famous photo of Elvis meeting Richard Nixon.

Thus the White House came up with a work-around: The president would deal directly with the obscure chairman of the board, a Swede named Carl-Henric Svanberg. For the meeting at the White House, Svanberg would be the leader of the BP delegation. Hayward and BP executives Bob Dudley and Lamar McKay, both American, would be allowed to tag along.

The meeting was scheduled for the morning of Wednesday, June 16. The administration wanted BP to create an escrow account into which it would pour billions of dollars to meet the claims of Gulf Coast residents, fishermen, shrimpers, hoteliers, and so on, who had been hurt by the spill. The White House named a number: $20 billion. That would be a big escrow account even by Big Oil standards. All this would have to be hashed out between the administration and BP prior to the actual sit-down at the

White House. In Washington, the hard decisions get made in advance. You don't want the president negotiating on the fly with BP executives—that's below his pay grade. The president isn't a political roughneck, he's the boss of the boss of the boss. And if he gets so much as a smudge of oil on his hands, someone screwed up.

BP understood the political dynamics: The White House needed to score a victory. BP could afford to let the White House do that, so long as it fit into a larger strategic objective for the company: limiting the liability. Such political victories for the administration would reduce the pressure on the company—they were like those rupture disks in the sixteen-inch casing of Macondo. Manage your pressure. A controlled burst is better than a blowout.

Even as the news spread of the imminent face-off in the White House, new information surfaced about BP's engineering decisions leading up to the April 20 explosion. On Monday the fourteenth, Henry Waxman, chairman of the House Energy and Commerce Committee, and Bart Stupak, chairman of the Subcommittee on Oversight and Investigations, sent a letter to Tony Hayward that quoted from the email traffic showing BP engineers making last-minute changes in plans on the Macondo well, worrying about the lack of centralizers, and pondering the costs of different procedures. They wrote:

> In spite of the well's difficulties, BP appears to have made multiple decisions for economic reasons that increased the danger of a catastrophic well failure. In several instances, these decisions appear to violate industry guidelines and were made despite warnings from BP's own personnel and its contractors. In effect, it appears that BP repeatedly chose risky procedures in order to reduce costs and save time and made minimal efforts to contain the added risk.

The letter received widespread play in the media and advanced the narrative of reckless engineering by BP.

Obama shot down to the gulf again on Monday the fourteenth. He flew home the next day. Thad Allen joined the president on the trip. On the way back, aboard Air Force One, Allen found a seat on a small couch in the passageway not far from the president's office. He and an Air Force steward struck up a conversation. Then the steward got up and left, and Barack Obama sat down next to Allen.

"So, Thad, how's it going?" Obama asked. This was not a casual chat, though. The president was going to go on national television that very night to address the nation about the oil spill. He needed to know exactly what was happening, directly from the government's point man. Allen said: "I need to take control." For one thing, he said, he needed to militarize the airspace. There had been eight near misses of aircraft over the spill, he told the president. Obama told the admiral he should do whatever he needed to do.

"There are no do-overs," Obama said.

"Mr. President, I understand that," Allen replied.

That night, Obama held his first prime-time address to the nation from the Oval Office. The president might well have gone through his entire term without such an address, preferring in general to downplay drama rather than crank it up. But this had become, for the moment, the biggest story in the country, overshadowing the wars overseas and the gasping economy. It had rattled people. It was scary in a peculiar way—*indeterminate, asymmetrical, anomalous*. The president had to say something. So he said that the government was in charge of this spill, that he'd gotten his Nobel Prize winner to ride to the rescue, and so forth:

> Because there has never been a leak this size at this depth, stopping it has tested the limits of human technology. That's why just after the rig sank, I assembled a team of our nation's best scientists and engineers to tackle this challenge—a team led by Dr. Steven Chu, a Nobel Prize–winning physicist and our nation's secretary of energy. Scientists at our national labs and experts from academia and other oil companies have also provided ideas and advice.
>
> As a result of these efforts, we've directed BP to mobilize additional equipment and technology. And in the coming weeks and days, these efforts should capture up to ninety percent of the oil leaking out of the well. This is until the company finishes drilling

> a relief well later in the summer that's expected to stop the leak completely.
>
> Already, this oil spill is the worst environmental disaster America has ever faced. And unlike an earthquake or a hurricane, it's not a single event that does its damage in a matter of minutes or days. The millions of gallons of oil that have spilled into the Gulf of Mexico are more like an epidemic, one that we will be fighting for months and even years.

It was unclear what Obama intended as the take-home message. Should people feel reassured that the "nation's best scientists and engineers" were on the job and that 90 percent of the leak should be captured somewhere down the road? Or should people feel sobered, and worried, that this was like an epidemic (!) that we'd be fighting "for months and even years"?

People wanted to hear that this would end. They wanted it to go away.

"Spring Rain," a pseudonymous commenter on an online *Washington Post* story, surely spoke for millions: *"God, please, please, please! Make it stop! Please God, make the oil flow into the gulf stop—I can't take it anymore!"*

Thad Allen woke abruptly at four o'clock the following morning. Within minutes he was at his computer, pounding out an email. He'd been inspired by his meeting with the president on Air Force One and by the president's address hours later. The gravity of the situation had always been clear, but the admiral was really feeling it now, and he wanted to do something new, different, bigger. His email, when printed, ran to five pages, single spaced. Allen said it was time to militarize the airspace in the northern gulf. They would have to invent response doctrine on the fly. They would increase the number of oil-skimming boats. They'd flood the zone on this oil spill.

"There is no such thing as doing too much in this response," he wrote in conclusion.

Hours later, the BP executives arrived at the White House. They had their apologies rehearsed. But they also had some strategic needs.

The administration demanded that BP put up $20 billion in an escrow account to deal with spill damages and claims; the figure was neither a ceiling nor a floor for what BP would ultimately have to pay. (The terms of the escrow account, when finally settled, were favorable to BP in a key respect: The collateral would come from the company's United States operations. This seemed to ensure that BP wouldn't be banned from drilling in the gulf.)

BP succeeded in getting the White House to retreat on the issue of compensating oil industry workers for lost wages due to the government deepwater drilling moratorium. BP would instead offer a one-time payment of $100 million to help the idle workers. "We made clear that we do not think this is a liability for the company. The president said he's concerned about those workers. He asked if there was something we could do as a voluntary gesture," recounted Jamie Gorelick, a former deputy attorney general in the Clinton administration who had been retained by the oil company and spearheaded the negotiations with the White House.

The BP team had one more, highly important request: The company desperately needed Obama to go before the cameras after the meeting and declare to the world that it was in the best interest of the gulf and its citizenry for BP to remain a strong and viable company. BP needed, in essence, for the president to say that he didn't want to see it go bankrupt.

Obama obliged.

"BP is a strong and viable company, and it is in all our interests that it remain so," he said, speaking very deliberately, when he appeared before reporters in the State Dining Room.

Obama had rescued BP. The company's troubles weren't about to go away, but BP had reached what amounted to a cease-fire with the US government. This felt doctrinal. The Obama administration was never going to be BP's ally, friend, soul mate; the Justice Department, after all, had launched civil and criminal probes of the company. The government would surely sue BP under the Clean Water Act. But the Obama administration prided itself on being pragmatic, and it knew that the effort to hold the company accountable for the spill would be stymied if BP went bankrupt. This was a twist on the old saying "If you owe the bank $100, you have a problem; if you owe the bank $100,000, the bank has a problem." BP's problems were the administration's problems. The company was "too big to fail," in a sense. Hence, a deal. Hayward later told the BBC that the meeting put BP and the government "in alignment," and that there was an "almost unspoken agreement" that the political rhetoric would be toned down.

The meeting gave all parties what they needed most. BP got some clarity about, and limits to, its long-term liability; Obama demonstrated leadership by securing a giant pot of money for the oil spill cleanup and economic recovery on the Gulf Coast.

Stepping outside the West Wing, to where the reporters and camera crews were standing on the ceremonial White House driveway, the BP executives did their best to look remorseful, apologetic, humbled, submissive—like dogs caught and severely punished for eating the crème brûlée before a dinner party.

Svanberg spoke for the team: "We have made it clear to the president that words are not enough. We understand that we will and we should be judged by our actions." He spoke in a thick Scandinavian accent. "I would like to take this opportunity to apologize to the American people on behalf of all the employees of BP, many of whom are living on the Gulf Coast."

All good. But then came a verbal blowout. Obama, he said, "cares about the small people. And we care about the small people."

The small people.

The emergency apology that came from BP later, in which Svanberg acknowledged speaking "clumsily," was not enough to head off a brief media kerfuffle about this man with the strange accent talking about "the small people" that he probably would be pleased to see shine his shoes or wax his limousine or whatever it is that the big people of the oil industry like the small people to do.

Never mind that English wasn't his first language and he didn't mean anything bad by it. In the summer of the spill, no one got any slack.

The next morning, Tony Hayward faced the firing squad. He had been asked to appear before the House Subcommittee on Oversight, part of the Energy and Commerce Committee. He was not given so much as a final cigarette.

Bart Stupak, the subcommittee chairman, used his introductory statement to hurl Hayward's gaffes (and Svanberg's, too) back in his face: "I hope we will hear honest, contrite, and substantive answers. Mr. Hayward, you owe it to all Americans. We are not 'small people' [*zing!*], but we wish to get our lives back [*zing!*] . . . For the Americans who lost their lives on the rig, their families may never get their lives back [*zing!*]. Mr. Hayward, I am sure you will get your life back [*zing!*], and with a golden parachute [*zing!*] back to England [*Eeeeeenglaaaand!*]. But we in America are left with the terrible consequences of BP's reckless disregard for safety."

And so it went. It was murderous. For more than an hour, Hayward had no opportunity to make his own statement. One lawmaker after another, with a single exception, lectured him on all that he and his awful company had done wrong. The protocol in Washington is that, before someone testifies in a hearing, each member of a committee is permitted a brief opening statement. So is any high-ranking congressman who happens to exercise his or her prerogative to sit in on the session. Lawmakers usually have a million things to do and attend these sessions fairly briefly, leaving their seats empty for the majority of the hearing. Not this time. This was a front-page event, potentially; this was going to be on the cable networks. Everyone showed up, hauling supertankers of disdain to dump on Hayward.

He sat in silence. He looked like he hadn't slept for a couple of days.

He could find only one friendly face in the crowd: Joe Barton, a Republican from Texas. Barton, seemingly unaware that BP had gotten pretty much what it needed out of the White House session, apologized to Hayward for how BP had been treated by Obama:

"I am ashamed of what happened in the White House yesterday. I think it is a tragedy of the first proportion that a private corporation can be subjected to what I would characterize as a shakedown—in this case, a twenty-billion-dollar shakedown . . ."

The Barton apology came early in the hearing and instantly became the story of the day, leading the wire stories and TV news reports. Barton later tried to qualify his statement, but it was too late, and the Barton Apology obscured the deeper truth that, over the course of many hours, the lawmakers beat Hayward to a pulp.

Hayward's best moments came early on when he said nothing and simply took the verbal pummeling. When he finally spoke, things got worse. His prepared statement included some contrition, and empathy for the people of the gulf. He said all the right things about cleaning up the gulf and never letting this happen again. But when he was questioned by the committee members, he defaulted to prepared responses that said little. When asked, repeatedly, about alleged errors by BP prior to the blowout, he said that if anyone traded safety for cost, there would be repercussions. He said it again and again and again. It was bland. He didn't name names or admit any specific wrongdoing. Hayward sounded supremely lawyered up and kind of . . . sleepy. He seemed like a man having such a miserable time that he had shut down all but a small portion of his brain, a kind of strategic zombification. He had no real information to share, nothing to tell people that they didn't already know. He had nothing to say except "Sorry." No one wanted to hear a weak apology: They hated him and his company and his oil spill and his fancy British accent far too much to accept one.

The very next day, Carl-Henric Svanberg let drop during an interview that Bob Dudley would be taking over the leadership of the BP unit handling the gulf spill response. Dudley had already been tapped to take over the response once the well had been plugged, but the well hadn't been plugged yet, and the Dudley move had been accelerated. Hayward flew home to England. That Saturday, no doubt still wrung out by the events in Washington, he attended a yacht race at the Isle of Wight. His personal fifty-two-foot yacht, *Bob,* participated in the popular J.P. Morgan Asset Management Round the Island Race. Once again, the BP public relations machine must have had sand in its gears. No one told Hayward that, in the United States, a yacht is the singular trademark of the idle, snooty rich. It's the equivalent of plucking a highball from a silver platter held by a servant named Barnstable. Hayward could not have made a worse choice for an activity had he decided to spend the weekend undergoing an exorcism. Most Americans can't even spell *yacht,* and they resented Tony Hayward—this *yachtsman*—telling them that they are the small people (even if it was that other guy, the Swede, who said it—but whatever!).

Hayward had become a joke. At a June 21 White House briefing, a reporter posed a question to deputy press secretary Bill Burton:

"Tony Hayward and his yachting habit, does the—the administration see that as just a problem with Tony Hayward or is that reflective of how BP has addressed this entire problem, do you think?" Burton's reply: "You know, look, if Tony Hayward wants to put a skimmer on that yacht and bring it down to the gulf, we'd be happy to have his help."

Within days, word came that Hayward would not be returning to the United States at all but rather would be traveling to Russia to attend to BP corporate interests completely unrelated to the gulf spill. The industry could not help but make the obvious joke: Tony Hayward was being sent to Siberia.

Days later, Dudley made a visit to Washington and, in the process, met with a small group of reporters at the offices of BP's public relations firm, the Brunswick Group. He came into the conference room and apologized for being a little bit late—said he'd just come from talking to Ken Salazar, the interior secretary. But he didn't have that running-late look. He seemed to be by nature an extraordinarily calm person, another BP astronaut, someone with exquisite blood pressure and a heart rate of fifty-two beats a minute. Compared to this new guy, Tony Hayward was a Mexican jumping bean. Dudley was neatly groomed, with wispy blond hair. He wore a crisp shirt with a spread collar and a carefully knotted purple striped tie. His voice is soft, his hands are big. He looked like a preternaturally cool customer, someone who could navigate the capital's ninety-five-degree summer heat and the glares of dyspeptic Cabinet secretaries without a scientifically measurable trace of sweat. This guy, he wouldn't wince if you detonated a firecracker next to his head. If you're BP, having an existential crisis, unsure of whether you'll even be in existence in a few days, naturally you turn things over to a man made of space-age polymers.

And the most important thing was: He said the right things. The corporation could have confidence that Dudley would be no more likely to produce a gaffe than grow a second head.

Dudley told us that he'd been in India when he was called to fly to Houston. He arrived May 1 with a small suitcase, and had been on the road ever since. He said the creation of the new BP operation he was heading, the Gulf Coast Restoration Organization, had been "somewhat accelerated" after the meetings at the White House. (BP would soon announce that Hayward was leaving the company and would be replaced by Dudley.) Dudley ran down all the things that BP was doing to kill the well and help people in the gulf. "I'm confident"—he rapped his knuckles on the wooden conference table—"that by the end of August we'll have that well killed."

But he also seemed realistic about the fact that BP wasn't going to have any friends in the meantime. BP was the villain of the story. People hated his company. They hated his entire industry, he said: "The oil industry has been an unpopular industry in the US for a long time, but it employs hundreds of thousands of people and pays billions of dollars in taxes." But he said he could understand the fury of the American people. He understood that this remained an open wound, this uncapped well: "It's not a 'spill.'" He said that as long as the plume existed, gushing away, the situation was fraught with "infinite uncertainty."

Explosives, Again

If only they could just blow up the well. Pound it. The idea would never go away; it had such visceral appeal. No more monkeying around with Macondo. Man has mastered the art of explosives, he has extended his muscularity to unthinkable dimensions. A species with a gift for creation has cultivated the power of destruction. Blowing up the well might seem cockamamie at first glance, and then, at second glance, even more cockamamie, but the idea kept popping up in the brainstorming sessions. More than two full months into the crisis, the Chu science team found itself once again thinking about the use of high explosives in the well.

Alex Slocum, the MIT professor and gizmologist, got the conversation going with a provocative early morning June 24 email to Chu, Hunter, and the rest of the now expanded science team, which included the president's science adviser, John Holdren. The email had the subject line "worst case contingencies." An excerpt:

> So our worst case scenario seems to be: Relief wells do not work (even with previously just emailed ball bearings/mudd mix idea),

> new structure on flex riser toooo cantilevered and production ships do not have capacity . . . . . . .hurricanes approaching (probably also dogs and cats living together . . . .[if you have not seen the original *Ghostbusters* movie I highly advise it]).
>
> Soooooo then its time to cross the streams . . . .
>
> We will have two very nice holes from the surface down to the hurt well—below the level of the rupture disks—
>
> Time to insert LOTS of very high explosives . . . . . . and detonate them at the intersection of the relief/Macondo bores . . .

In a subsequent email, Slocum made clear that this would be a backup plan in a worst-case scenario: "Methinks we need to seriously start these calculations (something only the DOE lab can do) and then tell BP this is the plan if the relief wells do not work."

Tom Hunter, who had spent two months riding herd on all these geniuses, responded to Slocum's email with what should have been a conversation-ending one-liner:

"It seems like a good opportunity to turn a potentially controllable situation into an uncontrollable one."

Minutes later, Marcia McNutt echoed Hunter:

"Alex—This technique (fracking) is used to increase the flow of wells."

So, bad idea. Obama wanted them to stop the flow of Macondo, not increase it.

Then Richard Garwin, the physics legend, weighed in, also with a negative assessment of the idea. Garwin went on at length as to why it didn't make sense. He wrote: "For weeks people in and out of the national labs have been discussing nuclear and non-nuclear explosive approaches to well closure." He cited a proposal from an outside scientist to use small nukes near the seabed to vitrify the rock. The scientist, in writing to Garwin, was dismissive of the hand-wringers who were reluctant to go nuclear: "I for one believe that—while there may have been in the past a big barrier to an 'ultralite' nuclear solution for searing and sealing the upper seafloor—this public negativity is now (and progressively so) mostly in the minds of politicians, bureaucrats, and the media." But Garwin, who knew a thing or two about nuclear weapons, what with having helped invent them, said that there were too many unknowns in the mix to permit an explosive solution.

Garwin wrote, "Desperation or exhaustion is not a strategy. When several approaches do not work, the next one is no more likely a priori."

Hunter seconded Garwin: "Well said. This is an idea that just won't go away."

Slocum wouldn't let it go. He wasn't suggesting a nuclear option, just a backup plan involving high explosives: "What I proposed was *not* desperation—it was entirely logical. Array of small nukes to vitrify . . . that's desperation."

Garwin, not famous for his sensitivity, recognized the bristling tone in Slocum's last email, and he prepared a response that would come to be mythic within the science team. By this point, the early morning email thread had extended past midnight, into the wee hours of June 25. At 2:18 a.m., Garwin wrote: "Dear Alex, I hope that I wasn't rude, but this is a topic I know a good deal about." He cited a National Academy of Sciences study that he was just completing that looked at the containment of underground explosions. He said he worked on a 2002 study on the same topic and wrote a 1975 paper on peaceful nuclear explosions. And then he wrote:

"Here, are 7 pages in Enrico Fermi's hand from my July 1950 Los Alamos Notebook—an early calculation of such containment."

Garwin had dug up a notebook from 1950 in which the great Fermi had handwritten some calculations on the propagation of shock waves from explosions. Garwin scanned the seven pages and attached them to the email.

That conclusively ended this latest iteration of the blow-up-the-well conversation. The rest of the science team was duly impressed. This was a Garwinian trump card. How do you argue with Enrico Fermi, the giant of physics, the man who ignited the first sustained nuclear chain reaction, discoverer of elements, mentor to the elite in a generation of physicists, and so on? It was as if Garwin had saved his Fermi notes—"in Enrico Fermi's hand"—for the last sixty years for precisely this moment. This was like the famous scene in *Annie Hall* when Alvy Singer (Woody Allen), waiting in line for a movie, gets in a debate about the theories of Marshall McLuhan and pulls McLuhan himself from behind a sign to yell at the idiot who doesn't know what he's talking about.

I happen to have Enrico Fermi right here . . .

A Disaster Area

Despite the $20 billion "shakedown," BP began to gain some financial stability, and the company's stock price no longer was in free-fall. Contravening all known rules of Macondo, things started to go right with the subsea response. Technology started working as advertised. The containment option actually began containing large amounts of oil. The first of the two relief wells, being drilled by the Development Driller III, another Transocean rig, was homing in on Macondo. A second relief well, started two weeks later, spiraled in from a different direction. It was now possible to imagine, if not yet discuss in polite company, that this gulf-poisoning, dream-haunting oil spill might actually come to an end one of these days.

Or was this delusional thinking? Few people in the general public expressed optimism about the course of events. The news media may have helped create the strong impression that any good news about Macondo was by definition misinformation. A CNN/Opinion Research poll released June 18 reported that half the country (48 percent, to be precise) thought the situation in the gulf was getting worse. Only 14 percent thought it was getting better. Half predicted that the gulf would *never* fully recover.

Matt Simmons, the doomsaying oil investor, told me one day that, notwithstanding his many bleak pronouncements, he'd actually been underestimating the extent of the horror. "This story is eighty times worse than I thought," he said, and went on to offer his own maverick theories for what was happening on the seafloor. He said the blowout preventer that we were looking at was no longer on top of Macondo but had been blown some distance from the well bore. He said the well bore was probably eroding, growing larger, that "it might be twenty or thirty feet in diameter." The upshot: "We're going to have to evacuate the gulf states."

Some scientists predicted a busier hurricane season than usual. The pessimists knew that tropical cyclones would be lining up in the steamy Atlantic like jetliners taxiing to the runway. The concept of an "oilcane" rocketed around the Internet.

"The oil spill is bad, but it could become much, much worse and soon," began one piece that ran on a number of blogs "If a hurricane's violent winds track over the spill, we could witness a natural and economic calamity that history has never recorded."

This dire talk formed the backdrop of BP's effort to build a better containment system for the gushing Macondo crude. The centerpiece of the new plan was the 3-ram capping stack, which could fit tightly on the chimney of the well. The new stack was something like a blowout preventer. It had multiple valves that could open or close, and three rams for shutting in the well. It would go right on top of the original, "legacy" BOP. This was the BOP-on-BOP idea that had been around since the early days of the response. In fact, the 3-ram capping stack, manufactured by Cameron—the same company that built the blowout preventer—had been available to BP since early May. The engineers hadn't planned to use it because they believed the top kill would terminate the well.

The switch from top hat to 3-ram capping stack would be much harder than any technical operation attempted so far. BP engineer Kent Wells reminded reporters, "This isn't something where you just snap the cap off and put the valve on. It's not that simple."

To those of us on the outside looking in, the subsea containment operation was quite bewildering. There were all these ships, rigs, top hats, cofferdams, free-standing risers, blowout preventers, and now this new piece of hardware, the 3-ram capping stack (or was it the 3-ram stacking cap??? Ramming stack? Wha . . . ?). An opportunity arose for a little more clarity: In late June the Coast Guard began taking reporters to the disaster site. All the talk at news conferences about what was being done, the progress on the relief wells, the containment efforts of the *Enterprise* and *Q4000,* could never quite convey the scale of the response, the intensity of the operation.

But now we'd be able to see it for ourselves.

* * *

I arrived at the BP heliport in Houma on the morning of June 26. A low-slung structure sprouted temporary buildings with air-conditioning units humming away in the searing heat. Inside, a sign on the wall, under the heading "BP's Commitment to Health, Safety and Environmental Performance," stated:

"Our goals are simply stated. No accidents, no harm to people, and no damage to the environment."

A small group of reporters and photographers had gathered for the flight to a rig at the source. A Coast Guard public information officer gave us a brief orientation. We watched a video about the rig's safety plan. No alcohol on board, no firearms, and smoking only in designated areas. No necklaces or earrings allowed. Steel-toed boots worn at all times. Safety glasses required outside accommodations area. The video explained certain protocols for ensuring heightened safety awareness, such as the START program: See, Think, Act, Reinforce, Track.

Clean clothes must be worn at all times in the accommodation area.

"No horseplay is allowed at any time in any area of the installation."

The helicopter flight to the rig took about an hour. The helicopter was unpressurized; we wore life vests and ear protectors. It's the same route the BP and Transocean executives took when they flew to the Deepwater Horizon on April 20. This time, though, you could see the dredges building the Jindal and Nungesser sand berms just beyond the bird's foot delta. And then the oil. Just beyond land, the first streamers appeared. In every direction, you could see long, stringy streamers of red-orange oil, looking like the remnants of a birthday party. The sheen on the water stretched as far as the eye could see, maybe forty miles, a slick with no end.

Finally, out the windshield of the helicopter, the disaster site came into view. On April 20 there would have been a single rig out here, with a workboat alongside and maybe a few pleasure craft tooling around. Now it looked like a city in the middle of nowhere, a Las Vegas of the sea. It was a sprawling complex of structures: some towering, some squat, most of them colored a dull gray, the gray of metal. Think glass-and-steel modernist architecture, minus the glass.

In the middle of all this were the two flares, the jets of burning hydrocarbons. One flare, burning oil, emerged from the side of the *Discoverer Enterprise,* a Transocean drill ship with a towering derrick. The second flare, burning both oil and gas, came from the Helix *Q4000,* a lumbering, rather humble-looking servicing rig. The fires were so intense that workboats had to douse the pipes to keep them from melting.

From miles away, the disaster site looked like a city on fire.

The scene had become a roaring industrial complex with more than sixty vessels, each dynamically positioned using GPS, pivoting around the focal point represented by the *Enterprise,* which was directly over Macondo. The two drilling rigs dominated the scene, each one as big as the Horizon, with twenty-five-story derricks in the center. The chopper nearly circled the site before setting down softly on the helipad of the Transocean Development Driller II. The rig was drilling the second, and backup, relief well.

Step out, there's no sign of pitch or roll. This thing floats? It's as steady as a parking lot. Peer over a handrail, and you can see, below the water, the white outline of the pontoons that keep the rig afloat. The rig is huge and complicated, seemingly built by someone adding random elements.

I kept looking across the water at the flares, the oil flare of the *Enterprise* and the oil-and-gas flare of the *Q4000.* The flames were invariant, never sputtering, never faltering. They were just this side of being too bright to gaze upon, and were borderline hypnotic. The *Q4000* and the *Enterprise* appeared to be almost close enough to reach with a well-tossed football, but no, they were about a half mile away. Distances on the open water are deceiving.

A crewman stood at a water cannon at the edge of the helipad. The cannon was aimed at the helicopter, ready to douse it in case of a conflagration. The crewman was wearing a protective suit that appeared capable of surviving a massive fireball. I did not find any of this particularly reassuring. I asked the man, John Antoine—who has that marvelous oil-patch title "roustabout"—if he was hot in the fire suit. "I'm trying to lose a couple pounds so when I get home I'll be a little trimmed out," he said.

We went down some steps to a suite of air-conditioned rooms. The cultural climate of the Deep South features two seasons: blazing and air-conditioned. One minute you're in the tropics, the next in the Arctic. We signed in and gathered in a room where we were given some orientation, evacuation instructions (our lifeboat would be the helicopter), snacks (apples, grapes, chips, Big Red cola perfectly iced down), and whatever protective gear we hadn't brought with us. Then the tour began, and we got to see and feel and hear this extraordinary apparatus called a Mobile Offshore Drilling Unit.

There were 173 people aboard, all of them men, as far as I could see. A rig boss said there were actually "about five" women on board, but I didn't see one. This is just about the last male workplace outside of the combat infantry. They were watching NASCAR races in the cinema the night before our visit; one guesses that the crew is not likely to request a screening of *He's Just Not That Into You.*

And it's a very hard place, literally. All the surfaces are metal. You walk on grates where you can see, far below, the oily blue of the Gulf of Mexico. (You might want to avoid looking down.) There are handrails everywhere, and metal stairs. The rig has multiple levels and, being rectangular, has no obvious front or back. The rig is loaded with computers, space-age technology, wireless communications, and robotics, and yet it is all in the service of making a hole—in this case, making a hole to kill another hole.

A simplistic formulation came to mind: Technology caused the oil spill, and technology would fix it.

A more cynical formulation: They haven't had a new idea for killing a well in decades. They're drilling a relief well just like they did at Ixtoc in 1979 and Santa Barbara in 1969. Never mind the computers and satellite technology and whatnot, this remained an old-fashioned response, a mud shot to the base of Macondo, just like the olden days.

We got to the drill shack—the "doghouse"—a caged room, air-conditioned, where a couple of beefy guys sat in swivel chairs and monitored video screens while talking into a microphone to crew members outside on the drill floor. They were focused on the operation at hand and didn't pay a lot of attention to the visitors. Another worker, sitting in the rear of the shack, seemed a bit surprised that the rig's task was worthy of so much attention.

"You came all the way from Washington?" he asked me.

Was it possible they didn't realize that the rest of the country was going completely bonkers over this oil spill? Would it be impolite to grab the guy by the front of his coveralls and say *"Make it stop! Make it stop! Make it stop!"*?

But no: They were doing their jobs, doing them by the book, carefully, skillfully, focusing on the task at hand, and not getting distracted by the braying of the outside world. A good rule in a crisis is, at the point of attack, keep the professionals in charge. This is the battle cry of competency. Don't let a crisis put you off your game. Don't rush, don't panic, don't deviate from best practices. And thus, to a striking degree (striking at least to me, coming from the world of *Make it stop!),* this looked like engineering as usual—only with a lot more supervisors looking over everyone's shoulder, and supervisors of supervisors, on up an implausibly long chain of command, until way up there in the command stratosphere you had Steve Chu reviewing pore pressures and fracture gradients.

"We want to get this thing done so bad it hurts," said Mitchell Bullock, one of the BP company men on the rig. "But you can only do it at a certain pace."

Chris Wokowsky, forty-nine, the offshore installation manager for Transocean, said of the workers, "We want them to look at this as drilling a well. We don't want them to be distracted."

The biggest guy in the drill shack sat in the driller's chair. He was Jeremy Marts, age thirty-one, and he said his title was simply "driller."

"It's just a process that takes time," Marts said. "It doesn't happen overnight. We're out here drilling a well. That's our focus. I wake up in the morning, my focus is to keep all these guys safe for the twelve hours I'm working."

Through the glass top of the drill shack, which is protected by steel grating, we could see 149 sections of drill pipe racked in the derrick, each pipe 122 feet long and weighing 34 pounds per foot—heavy-duty stuff. My college geology professor, Ken Deffeyes—famous as a proponent of the peak oil theory and as a character in the classic John McPhee book *Annals of the Former World*—has helped me understand the petroleum engineering industry, and offered an emailed explanation of what I was looking at in this particular moment:

> The standard length of drill pipe is 30 feet; your 122 feet is four joints of drill pipe, a "fourble." You have to learn to count the number of joints in a stand: single, grayhound, treble, fourble. The catwalk about three-quarters of the way up an old-fashioned

> steel derrick is a "fourble board." The thicker, threaded connectors between joints of drill pipe are "tool joints." The Ixtoc blowout happened because the shear rams in the blowout preventer happened to be opposite a tool joint.

It's a long shift, and, offshore, there is no such thing as a day off. The crew can play foosball in the rec room, or compete against one another in Wii. There's TV and Internet, and two workout rooms, one for cardio, one with weights. If it's Saturday or Tuesday, it's steak night. If it's Friday, you're eating seafood. Sunday is the day of fried chicken. There aren't a lot of surprises—this is a planned, managed environment.

We toured the bridge, just like the executives on the Horizon on April 20. A digital display showed the distances to all the other vessels. The *Enterprise* was 0.45 miles away; the *Q4000,* 0.63 miles; the Development Driller III, 0.47 miles. There were sixty-four ships within five miles.

There were also two ROVs straight down, on the gulf floor a mile below us. I ducked into the ROV shack and watched a man named Dean Miller operate an ROV stationed at the blowout preventer atop the relief well.

"We regularly see deep-sea rays, squid, octopus," Miller said. I asked him if he'd fly the ROV around a bit so that I could get a better view of the blowout preventer. The image was monochromatic, an eerie, flat blue-green. The BOP comes fully into view, but its scale is hard to discern. It must be big, but maybe it's only a foot tall—you can't tell.

Soon we returned to the chopper, our lifeboat. As the helicopter lifted off, the roars of the flares were drowned out by the thumping of the rotors. The visit had been brief, two and a half hours, and arguably a little superficial. But it was also revelatory. We had seen the source, ground zero of the catastrophe that had transfixed the country. We had gotten a taste of the offshore petroleum industry, which normally is hidden from view. The Development Driller II had more people, and more hardware, and was of greater economic significance, than many of the towns that dot a typical American road map, and yet it was not on any map. Nor could it be, since it was a mobile drilling unit. It was kind of like the International Space Station, something in its own realm, never fully appreciated, yet an engineering masterpiece.

Anyone visiting a rig like this would be impressed by the safety rules, the emphasis on wearing the right gear, the firm instructions to keep on the lookout for anything that might be risky. But these were matters of personal safety. My *Washington Post* colleague Steve Mufson has written that BP's safety culture was focused on the individual rather than on the systemic processes of drilling oil wells. Mufson made a trenchant point: Safety is more than wearing the right eyewear and footwear.

The men and women on the Deepwater Horizon were immersed in safety rules, safety measures, safety protocols, and operated in an environment of handrails, warning labels, lifeboats, life rafts. They trained for evacuations, trained for blowouts, trained for every kind of emergency they could imagine. And it wasn't enough.

As the helicopter circled the disaster site and began to fly away, I stared at the astonishing spectacle, the pinwheeling array of rigs and ships and workboats, the blinding flares of oil and gas, the seemingly chaotic but carefully choreographed complex of hardware—a flotilla never before seen on the planet. The sea surface was scribbled with a sheen, flowing in two separate ribbons from the *Enterprise,* heading generally north, toward Louisiana. What was it that Doug Suttles had written to Rear Admiral Watson? "Several hundred people are working in a confined space with live hydrocarbons on up to four vessels. This is significantly beyond both BP and industry practice."

No amount of safety rules, hard hats, earplugs, gloves, or steel-toed boots would drain the danger from the well-killing effort, or from the original endeavor, the puncturing of the earth, the quest for fuel, the raiding of the inner planet for its combustible compounds.

Calamity opens the eyes. This is the technological species trying to heal a wound of its own making. We need the energy, and look what we do, how deep we go, how willing we are to dig up the fossil fuels and burn them even if doing so transforms the atmosphere and gives our descendants a different planet from the one we were born into. This amounts to a massive geoengineering project that in the global sense is without serious supervision or forethought. No one voted for this, except to the extent that they voted with their feet, mashing gas pedals. For a century and a half, we have drilled for oil, but now we're penetrating to astonishing depths, going everywhere, devouring unimaginably remote reservoirs.

From Tolkien, we recall: "You fear to go into those mines. The Dwarves dug too greedily and too deep. You know what they awoke in the darkness of Khazad-dum . . . shadow and flame."

There are countless myths of hubris punished as men and gods play with the secret knowledge of the universe. Prometheus steals fire from Zeus and soon faces an agonizing eternity: bound to a rock, his liver pecked by a giant eagle (punishments being much more creative in those days). All technology is fundamentally hubristic. This is our chosen path, and in the main it has been a noble journey, fraught with benefits, extending our life spans, easing our pains, healing our wounds. But only a fool would assume that there will not be grim surprises along the way, feedback mechanisms, unintended consequences of enormous scale—Black Swans.

It's just a hole in the earth, a hole at the bottom of the sea. And yet such penetration of the planetary skin had always been—prior to the coming of our species—natural, and catastrophic, as was the case just south of here, off the Yucatan Peninsula, when a mountain-sized object crashed into the Earth 65 million years ago, an event blamed by many scientists for turning out the lights on the dinosaurs.

Humans, collectively, are akin to an asteroid impact, but let that be a longer conversation on another day. The well gushes, the plume rages, this is the moment to stay focused. This problem must be solved. The bigger picture can wait.

But it's true: We're playing with fire.

Chapter 11

Integrity Test

July arrived with high hopes and great anxiety. The time had come to switch caps on the well, build out the elaborate containment system, and attempt to capture all of the oil from Macondo. All of those involved in the subsea response had already been in battle rhythm for two months, but now they had to take it to a higher level still, mentally and metabolically at the very limit of human capacity.

Marcia McNutt, who as head of the US Geological Survey had dealt with the Haiti earthquake in January 2010 and, three months later, the Iceland volcano before getting sucked into the oil well blowout, could only dream of being back in her spacious office at the agency's headquarters in Reston, Virginia. She could only imagine riding her horse in the pastoral Piedmont in the shadow of the Blue Ridge. She was still ensconced in her windowless office in the Houston headquarters of an oil company that, with its cowboy culture, struck her as a throwback to an earlier era.

She could handle the bluster and swagger of the cowboys with no problem. She could talk the language of the oil patch, or delve into the geophysics of pore pressures, or bring out some of her knowledge from many years exploring the deep sea with ROVs or patrolling shallower water in scuba gear. Maybe they'd like to hear how she dove in a submersible nearly two miles deep on the flank of the East Pacific Rise. Cowboy culture? She could outride any man in the place if you're talking about actually getting on a horse. It is against that backdrop that McNutt likes to tell the story of the *Cosmopolitan*. One day she went up to the eighteenth floor to meet with the science team, and noticed, on the conference room table, a glossy magazine, *Cosmo*. She knew instantly that there must be another woman on the premises. There was, indeed—a top staffer from DOE had just showed up in Houston. McNutt had been so caught up in the daily grind that she had forgotten how much she missed female companionship, how much she'd been suffering from Man Overload. She spent her lunch hungrily reading the magazine from front to back.

"I suddenly looked at my sensible black sandals, my one pair of shoes that I had in Houston. I wanted to burn them. I wanted a pair of kitten heels. But no, I had to stick to my sensible sandals."

A colleague, Walter Mooney, walked in, saw her reading a magazine, and asked, "Are there any good carburetors in there?"

Steve Chu by this time had off-loaded many of his normal duties as energy secretary, postponed some department initiatives, cleared his desk of distractions, and focused his energy and his intellect on this unplugged hole at the bottom of the sea.

"We really got into the nitty-gritty of this," he said. "You go from seventy hours a week to eighty to ninety hours a week."

Somewhere in the mix, he oversaw the publication of science papers based on research he did before he joined the administration. On July 7 the prestigious British science journal *Nature* published a study by Chu and colleagues titled "Subnanometre Single-Molecule Localization Registration and Distance Measurements." It was Chu's second paper in *Nature* in 2010. This new paper, of which Chu was the lead author, described a way to see things that are very small. He and his colleagues had discovered a technique for seeing objects only half a nanometer across (a human hair is between 50,000 and 100,000 nanometers thick). Such objects are smaller than the wavelength of light, and thus are normally impossible to resolve with an optical telescope. Chu and his fellow scientists discovered a work-around, one that involved tiny fluorescent markers (presumably not something you can buy at the hardware store). *Nature* said that Chu's research "will be important for revealing the detailed workings of biological molecules, and may also find application in other fields that rely on precision imaging"—including astronomy, the study of things that are very large.

Chu, however, could not bask in the glow from his new *Nature* paper, what with this Macondo problem and other duties weighing him down. When someone asked him to comment about the new discovery, Chu demurred: "I don't have time. I live in several parallel universes, and occasionally they cross."

The Macondo universe was about to be transformed dramatically. The agent of change would be something called the integrity test.

The integrity test was made possible by the 3-ram capping stack, or sealing cap, that BP planned to put on the well if it got final government approval. The new cap had three outlets: a main vent up top, a choke line and a kill line. The main vent and the kill line could be opened and closed with a valve in an all-or-nothing fashion. Open, close. The choke line, however—as the name implies—could be closed incrementally. It could slowly choke the flow, until finally the well was shut in. During this procedure, the pressure in the new cap would increase dramatically. The engineers and scientists would study the pressure readings for indications that the well had integrity. If it didn't, the pressure would not rise as high as it normally would. Or the pressure could rise and then fall again suddenly as hydrocarbons exploded sideways from the casing deep in the earth—the much-dreaded underground blowout scenario. Then they'd quickly reopen the well.

So this would be framed as a test, an integrity test. In all public statements, BP and the government suggested that this was part and parcel of an oil-containment strategy, nothing more. If, for example, a hurricane blew in, and the surface vessels had to evacuate, the engineers could first shut in the well—presuming that it had passed the integrity test with an excellent score. When the storm passed, the vessels would return and resume collecting oil.

Somewhere along the line, assumptions had inverted. Just weeks earlier, BP had circulated the theory that the top kill had failed because the well was damaged. The reigning theory had been that this well could not be killed from the top because if you cranked it closed, the whole thing would go *kablooey!* at depth. Now people were talking about shutting in the well for a couple of days, maybe longer. The logic seemed to have evolved rather dramatically.

On the afternoon of July 6, Kate Baker, a geoscientist working under contract for BP, sent an email to Tom Hunter in which she discussed the company's plans for the new sealing cap. A few minutes later, she sent Hunter a follow-up:

"Tom, there is one important point I forgot to add. If the well integrity test indicates integrity, it could be possible to leave the well shut in until the kill team was ready . . . this test could give access to a scenario with no oil to the sea and no collection between 17 and 22 July. We would have to reopen the well for the kill, however."

Baker did not suggest that the integrity test by itself could end the spill—only that it could be extended, potentially, until the relief well intercepted Macondo at depth. There were a lot of storyboards in play, many dates in the mix, and the relief well seemed on track for a potential bottom kill sometime around the twenty-second of July, somewhat ahead of schedule. The relief well was always seen by everyone as the solution to the crisis.

The idea of shutting in the well—even temporarily—created a public relations dilemma for the government. No one wanted to raise expectations that might later get dashed. Thad Allen, for one, was militant about the need to underpromise and overdeliver. Plus, everyone was still feeling snakebitten by all the things that had gone wrong. The integrity test was a high-risk, high-reward maneuver. If BP tended to focus on the high reward, the government scientists felt it was their special duty to pay attention to the high risk. This could blow out the well at depth. This could trigger the Oilmageddon scenario of multiple leaks at the seafloor and Macondo bleeding out entirely. Explain *that* to the public. The engineers and scientists would have to be on high alert for the faintest sign of hydrofracking: of oil and gas breaching the casing and freelancing up through the rock. They would need to scrutinize the geology with every instrument they had, and they knew that what they saw would likely be enigmatic, open to interpretation. This was going to be an excruciating period.

And herein lay a timeless lesson: In a technological crisis, hardware isn't enough. The experts need more than caps, stacks, hats, plugs, slugs, seals, stents, and grommets. They need information. They need analytical tools, some way to model the unseen, some technique for anticipating future events and crafting contingency plans. They have to think really hard.

In the days leading up to the integrity test, the structure of the response—this bizarre coupling of government and industry—would be put to its own test of integrity. The public-private machinery of the response had never hummed along sweetly, but now, in crunch time, it threatened to make a god-awful racket. The BP engineers and government scientists needed to find a way to agree on a workable battle plan that countenanced every conceivable outcome. If it went bust, if there were an underground blowout, if the world beheld a spectacle of supreme horror as Macondo bled out into the gulf, then at least the people responsible could say (as they cleaned out their desks and began thinking of where they could find employment) that they'd made a carefully calculated decision.

BP spared no effort: It had contingency plans, avenues of retreat, and multiple redundancies. In addition to the new sealing cap, there were other top hats in the mix: top hat no. 6, top hat no. 7, top hat no. 8, top hat no. 10, all ready to be deployed in a crunch if the capping stack couldn't be seated properly.

As BP engineer Kent Wells put it, "We always have backups for our backups." Wells's "technical briefings" became more frequent. He starred in videos published on the BP website, and he held regular teleconferences with reporters. As the big day approached in which the new cap would go on, Wells offered more granularity in his description of the operation. He would say things like this:

> The way the sealing of the flange goes together, the flange is just not going to immediately come off. So we've devised a tool called our flange overshot removal tool, and we will come down on drill pipe from the drill ship *Inspiration,* and we'll put this overshot over the top of the flange. The gripping mechanism will get a hold of it, and we will pull from the *Inspiration* at the surface through the drill pipe and look to pull the flange off . . . Now, as always, we have a backup plan just in case that doesn't work. In the bottom

> middle [Wells is showing a slide at this point], you'll see the flange splitting tool, and we've actually devised another tool that would come in and, through hydraulic rams, force wedges between the two flange pieces and actually force them apart.

So for gearheads, this was totally cool and new and different and dramatic. This was hardware porn. The flange overshot removal tool! Who wouldn't want one of those?

Next came the question of timing. One day in early July, while he was visiting Houston, Thad Allen found himself in a long conversation in the war room hallway with BP's Andy Inglis.

"You know, the weather's pretty good," Inglis said offhandedly. "It's going to be good for a while."

A weather window . . .

Maybe it's time to rethink the plan, Allen said to himself. The current schedule called for sequential operations, very orderly, starting with ramping up containment capacity. They had wanted to get a new ship, the Helix *Producer,* attached to the well and "producing" oil. The vessel ought to be capable of boosting the containment figures by 20,000 barrels a day. Only after that operation was up and running would they remove the top hat, land the 3-ram capping stack, and conduct the integrity test.

Thad Allen now suggested an accelerated pace: BP should hook up the Helix *Producer* and switch the cap on the well simultaneously. That would require only eight days of calm weather, as opposed to twelve days for the sequential operation.

BP flinched. The company executives reminded the government, once again, that this was already an extraordinarily complex, hazardous operation; that the source was already teeming with vessels, that the engineers had their hands full with the SimOps, and that trying to do two procedures concurrently created more opportunities for collisions, fires, ROV lines getting tangled, and so on.

Allen double-checked the weather, writing to NOAA chief Jane Lubchenco early on the morning of July 7:

"Jane, BP meteorologist talking with your folks believes there will be a lull between tropical waves 16 and 19 as 17 veers northerly and that will produce a window to replace the cap. Can you be prepared to discuss at principals this evening?" ("Principals" was the five o'clock conference call, one usually crowded with Cabinet secretaries, their deputies, and White House officials such as energy czar Carol Browner.)

NOAA dutifully put its finger in the air and replied, yes, the forecast looked good, particularly given that this was July and they were talking about the Gulf of Mexico. Allen pushed for the accelerated schedule. BP agreed.

On July 8 Allen held a teleconference with the news media and did not mention that the 3-ram capping stack might actually shut in the well. He said nothing of an integrity test. He described the new cap in the usual way, as part of the expanded oil-containment strategy that offered greater flexibility in the event of a hurricane.

Finally, on July 9 Allen revealed for the first time what the government and BP were thinking. The news slipped into his daily briefing under cover of verbiage.

"We think this weather window presents a significant opportunity for us to accelerate the process of capping—shutting down the well from the top and increasing the prospects for being able to kill the well from below through the relief wells," he said.

Shutting down the well?

". . . once we will put a manifold or a valve system on top, that will allow us to basically shut in the well. . . . Our first goal would be to shut—we don't call it cap it, we call it shut the well in. In other words, close all the means of oil to escape."

Wait a minute: Did he say they could stop the leak? He seemed to be saying they could stop the leak. Indeed, he seemed to be quite explicit about it. Perhaps, though, everyone had misheard. At this point, the news media were not particularly attuned to the possibility that good news could infiltrate the prevailing narrative of misery and horror. The news found no purchase in our callused brains. The possible shutting in of the well generated few headlines.

Kent Wells didn't clarify much. On July 10, shortly after the cap-switching operation had begun, he obfuscated during a press briefing, at one point making a rather garbled statement—one that seemed to have misplaced a verb—about how the well "will get shut in, and then based on what pressure measurements we get, how it's opened up and what further steps are taken at that point."

What was he saying?

He later made another pass at it: "We will be doing this shut-in test that I have talked about, and at that point, decisions will be made on what goes forward in terms of longer-term containment, other options, et cetera."

Other options, et cetera . . .

Finally, he was asked directly by Mark Seibel of McClatchy Newspapers if the operation could leave the well permanently shut in:

"I'm curious: Once you begin the shut-in test, is it possible that you'll find the correct pressure readings that you're looking for and just simply seal the well at that point?"

Wells replied, "I think it's important that we go through the test procedure that we're currently designing, and that based on that information, we'll make the right decision going forward, and of course our priorities are always to minimize pollution."

We're going to choke the damn thing, and our long national nightmare will be over, you follow me?

No, he didn't say that, he couldn't say that. Everything had to be hedged, fuzzed, obscured in the service of expectations management.

Though perhaps slow on the uptake, the news media now understood that this change of caps in the well was far more than an incremental maneuver. That dawning awareness brought more second-guessing. Why now? We'd heard for so long that you couldn't kill the well from the top, that the myriad suggestions from ordinary citizens that involved plugs and caps and stents and slugs and clamps and binders and kinkers and chompers and whatnot were useless because of the damaged casing in the well and the danger that sealing the well would lead to an underground blowout. When did the wind reverse direction? Why couldn't this have been done sooner?

Doug Suttles, talking to reporters in a conference call, contended that the decision to go to the new cap came only after gaining information via other operations, such as the top kill. The engineers and scientists had to gain confidence that the new cap wouldn't make the situation worse. "The problem is, I've had to take these steps to learn the things

I've learned," Suttles said. "Without taking those steps, it's unlikely that I would have known what I know now."

Hardware isn't enough.

The Integrity Test

The complex operation began the morning of July 10 when the *Discoverer Enterprise* yanked top hat no. 4 off the well. Macondo gushed freely again. The ugly oil volcano of early June was back, spewing from the sheared-off riser.

Now came the procedure that had made insomniacs of the Houston engineers. The ROV pilots and technicians needed to use a jumbo wrench—wielded at the end of the fine manipulator arm—to remove the stub of the riser, known as the flange. It was attached to the top of the lower marine riser package (the connection point on top of the blowout preventer) by six huge bolts. If the ROVs couldn't remove the bolts, the flange would be stuck there, and the BP engineers would have to switch to hardware with a lower pressure rating. That could foul the entire plan, and it might be impossible to shut in the well—nixing the integrity test altogether. Hundreds of engineers, scientists, technicians, oil company executives, and politicians spent all afternoon and all evening nervously watching what might have been the most portentous bolt-removal operation in the history of technology. One bolt . . . two bolts . . . three bolts . . .

The last bolt came off at midnight. At three in the morning, BP's engineers and technicians dropped the overshot tool onto the flange and then lifted it off the blowout preventer. That exposed the flex joint atop the lower marine riser package. The flex joint had had its own issues: It had been pulled off the vertical when the disabled Horizon drifted during the fire on the night of April 20. All that force had tilted the flex joint 3 degrees, and so before any of this deep-sea plumbing could be done, they had had to use hydraulic jacks to straighten it out.

Engineers had assumed that when the flange came off, they would see two sections of drill pipe sticking up side by side in the flex joint. They'd seen these paired sections of drill pipe when they cut the riser. Why there were two drill pipe sections side by side inside the riser had baffled everyone. But whatever: The engineers had come up with multiple plans. They would lash the two pipes together with a specially designed strap, then drop a "mule shoe" over the both of them, in advance of landing the piece of hardware.

Except that the critical moment arrived, and there was only one drill pipe. Go figure. No pipe lashing needed. Onward!

BP deployed the twelve-foot-high, fifteen-thousand-pound Flange Connector Transition Spool. It was the hardware equivalent of an electrical adaptor: three prong to two prong. The ROVs then replaced the six bolts they'd removed earlier, tightening the spool in place. Down came the 3-ram capping stack. The ROVs guided it as it slid perfectly onto the spool and became, suddenly, the top piece of hardware on the well. Macondo now spewed from a perforated pipe jutting from the top of the capping stack. Check the clock: 6:20 p.m. (CDT), July 12. Day 84 of the crisis, and now the well had a tight-fitting cap. The ROVs removed the perforated pipe atop the stack; Macondo flowed through a round vent, the plume going full blast.

On Tuesday morning, July 13, BP made final preparations for the integrity test. All it had to do was turn the valve on the choke line, and the hideous oil-geyser would go away—at least temporarily. The operation had a scheduled start time of noon.

Steve Chu called time out.

Chu and his ace science team leader, Tom Hunter, had converged on Houston on July 12. ("It would be comforting to the Nation if you were there," Salazar had written Chu.) On the morning of July 13, just as BP prepared to choke the well, the science team had an anxiety-inducing conversation with engineers from several major oil companies. The outsiders talked about the many things that might go wrong if and when BP shut in the well. They discussed a scenario in which gas would surge up through the area immediately beneath the BOP, causing the gulf floor to liquefy.

"I've seen the entire ground around the BOP go plastic," Chu recalled one industry expert telling him, "and the BOP disappears beneath the surface."

The undersea quicksand scenario. Highly credible experts were talking about the possibility of the seafloor gobbling up the blowout preventer as if it were a Pop-Tart.

"We have a situation here," Chu informed Thad Allen.

To the BP executives, Chu said, "Give us a day."

It was not a request. Here was a moment when all the bureaucratic complexity and political posturing and awkward negotiating that resulted from the National Contingency Plan and its Unified Command structure and the ungainly arrangement of authority and responsibility between the government and the private sector finally slammed—smashed—into an unyielding reality: The government could say stop.

Some of the BP engineers were furious. The cap was on the well! They could turn off Macondo as if it were a faucet! They'd worked for nearly three months on this thing and now were a few turns of a valve from choking the well and turning the corner on this whole miserable situation—and Chu wants to double-check his math?

Military commanders deride soldiers who are reluctant to close with the enemy. Was this physicist quailing at the very brink of success?

Chu recalled telling the BP executives, "We want to make sure something catastrophic can't happen. You should support this. If the whole reservoir dumps, it is going to be awful." The culture clash between the government scientists and the BP engineers resounded through the Houston war room. Everyone had the same goal in mind, ultimately, because they had a powerful common enemy, but they were not truly allies at core. They had different interests, different ways of looking at the world, and were subject to different political and pecuniary considerations.

The public had no inkling of the backstage drama. In the newsroom, we studied the spillcam footage, waiting, waiting, wondering when they were going to shut this thing down. What were they waiting for? Finally, late in the evening of the thirteenth, Admiral Allen issued a statement: The integrity test had been delayed twenty-four hours. He said there had been some discussions, and "we decided that the process may benefit from additional analysis."

* * *

The government scientists had marveled at BP's startling change in attitude toward the likelihood that Macondo could be shut in safely. Just a month earlier, BP seemed to be pushing the notion that the rupture disks had, in fact, ruptured in the initial blowout, and that mud had been lost into the formations during the top kill. Now BP argued that the well probably had integrity, and even if it didn't, an underground blowout wouldn't be the end of the world.

Under one scenario, the rupture disks were providentially placed in rock formations of porous sandstones. There was much discussion of what were called the M110 sands, one of the layers in the geological formation. The M110 sands could supposedly act like a sponge, absorbing huge quantities of hydrocarbons leaking from the well. The gas and oil would naturally seek to migrate upward, toward the seafloor, but the formations would arrest that migration. Sensors would detect the migration of gas, and BP could reopen the capping stack. The new fissures in the rock would then collapse again, no harm done. It was a lovely concept: the self-healing underground blowout.

The government scientists went into overdrive trying to determine if these providential sands existed, and if so, would they plastically deform in the desired manner and arrest the hydrocarbons and then magically heal. But the government could not find sufficient evidence that such sands existed in the right location along the well bore. They might be there, or they might not be there—and that wasn't good enough to support a geology-will-save-us strategy. The precautionary principle had to apply.

Still, Chu and his fellow scientists were not ready to nix the integrity test. Instead he came up with what seemed like a reasonable solution: BP could go ahead with the test, but the company would have to intensify the monitoring of the well, in keeping with the government's recurring craving for more data, more diagnostics. At the government's insistence, a scientific ship, the *Geco Topaz,* had already conducted a seismic run over the well site—obtaining a map, in effect, of what the geological formation looked like prior to the test—and under Chu's latest plan, BP had to permit daily seismic runs by the *Topaz* after the test began. The ship would look for signs of gas moving in the formation (or, as the scientists put it, "phase reversals," "increased amplitudes," "velocity pulldown," and "acoustic disruptions"). McNutt said gas would be so reflective of seismic waves that it would look like a polar bear on a black sand beach. A NOAA ship, the *Pisces,* would also do acoustic monitoring, looking for gas bubbles in the water column. Finally ROVs would examine the seafloor, scouting for gas bubbles coming up through the muck.

The protocol for the test, as approved by the Unified Command, was framed around three scenarios. The gist was, the more pressure in the well, the longer they could keep Macondo shut in. If, when the well was choked at the top, the pressure in the stack did not rise above 6,000 psi, that would be a powerful signal that the well lacked integrity and that hydrocarbons were leaking down there somewhere in the formations. In such a case, the test would be terminated after no more than six hours. If the pressure rose to between 6,000 and 7,500 psi, that would be regarded as ambiguous. The test could continue for twenty-four hours. If the pressure kept rising, to 7,500 psi or higher, that would be a very good sign that the well wasn't damaged, and the test would be allowed to continue for forty-eight hours. After forty-eight hours, presumably, the well would be reopened.

That was just a guideline, though. The government could cut short the test and reopen the well at any moment. There would be high-level discussions every six hours to decide whether to continue with the test or reopen the well.

Richard Garwin didn't like the idea of the integrity test beyond a very brief initial shut-in. Too dangerous, he thought.

Writing to his colleagues, Garwin asked, "[A]re we relying on seismic and ROV Sonar to detect 'incipient broach to seabed?' How much good does it do to detect incipient broach? Our purpose is at all costs to *prevent* broach. Unless the damage can be prevented from continuing to broach, we should not shut in the well, beyond a brief initial pressure test."

But the decision had been made; the integrity test would go forward.

Gremlins delayed the test slightly. The choke line on the new stack sprang a leak and had to be hauled back to the surface and replaced with a spare choke line (timeless lesson: always carry a backup choke line).

While BP switched things around, the government obtained a happy piece of data. During the many hours it took to replace the choke line, BP diverted the flow of Macondo to the kill line of the new stack. Macondo surged from a single three-inch pipe. But about nine thousand barrels a day continued to be siphoned out of the well by the all-purpose Helix *Q4000,* which was still hooked up to the other kill line—the one attached to the old blowout preventer. Then, just as the integrity test was about to get underway, the *Q4000,* as planned, shut down its containment operation. The pressure in the kill line of the new stack jumped about 200 psi as all of the Macondo hydrocarbons now flowed through that narrow pipe. This presented Tom Hunter with a way to measure the flow of Macondo. He knew the *Q4000* flow rate, knew the pressure in the kill line before the *Q4000* went off-line, knew the pressure afterward, and knew the defining equations for flow through a valve. It was relatively easy to calculate Macondo's flow. The answer he came up with: 56,500 barrels a day. He sent his analysis to the rest of the science team under the title of "Fluids 101." The government science team, using additional data, refined the calculation into what became the government's best, and most precise, estimate of the flow of Macondo just prior to the integrity test: 53,000 barrels a day.

Early in the afternoon of July 15, Day 87 of the Deepwater Horizon crisis, the BP engineers and government scientists crowded into the Hive to watch the most significant subsea operation since the top kill. They looked at the ROV feeds from the gulf. The submersibles appeared to be swarming the new capping stack, patrolling the base of the blowout preventer, staring at the gulf floor. One ROV had latched onto the 3-ram capping stack and was prepared, when signaled by its human masters, to turn the choke. The most significant image of all presented itself on the left side of the room: pressure readings in the cap. This was all about the numbers. They wanted to see the pressure rise above 7,500 psi.

First came the coms check. The Houston engineers and the engineers and technicians out in the gulf had been linked since April by a conference call line. The technician handling the communications called out to each of the vessels, making sure they could hear what was happening.

Now came the moment to close the valve.

And then a strange woman's voice blared in the Hive.

"Hello. This is the Verizon operator . . ."

Whaa . . . ?

"I have noticed that this conference line has been open since late April 2010."

The Verizon operator was under the impression, evidently, that someone had mistakenly forgotten to close an old conference call line.

"Would it be all right if I closed this conference line?" she asked.

"Nooooooooooooooooooooo!" screamed everyone in Houston and the Gulf of Mexico.

"I'm sorry! I'm sorry!" the operator said.

The Hive technician informed the operator that they were in the middle of a very important situation, the operator hung up, and everyone could breathe again.

The integrity test began. In the gulf, the ROV's manipulator arm turned the handle on the choke line. Half a turn, stop.

At each stop, the people in the Hive watched the pressure. It would build a little, then reach equilibrium.

It took ten turns. On the final turn, at precisely 2:23:47 p.m. (CDT), the plume vanished.

The lights came up in the Hive. The lights almost never came up in the Hive, so this was a little bit startling. The Hive, with its multiple feeds from the ROVs, had been the ultimate bat-cave, inhabited by electronics and silhouetted figures wearing headsets, a place where voices came from the ether. But now there was light, and everyone looked around, shook hands, patted one another on the back, and permitted themselves a moment—just a second or two—of celebration, of congratulations, if not anything so decadent as a feeling of relief.

Because it wasn't over.

The hardest part would come now, in a way: the monitoring of the well, the scanning for leaks, the seismic surveys, the tough decisions about whether to keep it shut in or open it back up. No champagne yet.

But still, just look at the video feed.

The plume was gone.

They made it stop!

"There May Be Ominous Signs"

Now it was up to the well to hold, to prove it had integrity. At the start of the integrity test, gauges showed about 3,000 psi in the blowout preventer. The pressure rose quickly as the well was shut in. The scientists studied the pressure curve.

The earth spoke to them in the language of mathematics. They could not see anything directly, but they could try to interpret the unseen events, letting the numbers tell the story the same way that a facial expression conveys a hidden thought. To call this "remote sensing" would be an understatement. How could you, from a building in Houston, know with confidence what was happening below the gulf floor, miles down, in the fused sediments of a planet still hot beneath the collar after all these billions of years? You couldn't simply drill a test hole and drop some instruments to see what was down there. The relief wells had taken months to drill. The formations were known entirely through echoes of sorts—reflections of sound waves as they hit different layers of rock and sand and hydrocarbons. But that's not quite like taking a Polaroid of the buried sandstones.

Things were open to interpretation.

They did not get the numbers they wanted. The pressure quickly rose to about 6,600 psi as the ROV shut in the well, and then the pressure barely crept upward after that. It never dropped abruptly—a good sign, as that would indicate a big *foom!* down below. But neither did it show any sign of getting anywhere near 7,500 psi. This was decidedly ambiguous.

"It landed right in the middle of no-man's-land," Steve Chu said.

I spoke to Tom Hunter as the scientists were looking at the pressure curve.

"If it were a lot higher, it would be an easier decision to make," he explained. Hunter predicted that the pressure would eventually creep up to about 7,000 psi, but that still wasn't what the scientists had hoped for. For the moment, it hadn't even reached 6,700.

They now had to figure out why a reservoir that had been measured at around 12,000 psi was creating a shut-in pressure of less than 6,700 psi in the new capping stack. A fairly simple calculation, based on the depth of the well, showed that the difference in pressure between the reservoir and the wellhead should be something on the order of 3,000 psi. Had the shut-in pressure come anywhere near 9,000 psi, everyone would have been able to relax: That would unambiguously indicate no leakage down in the well. But now, with this much lower shut-in pressure, many of the government scientists feared that the hydrocarbons were surging into the formation, and could soon erupt through the floor of the gulf—the dreaded Oilmageddon scenario.

BP, however, seized on a benign explanation: that the reservoir had depleted—it had gushed for so long that it was running out of gas, so to speak. BP geologists and engineers said the pressure curve matched what they would expect from a well with integrity. The shape of the curve shouted "intact well!" To BP, it wasn't ambiguous at all. There were no strange jogs in the curve, no sudden drops in pressure, nothing but a classic shut-in pressure curve.

Those were two very different interpretations.

To this point, BP and the government had tried to be civil and constructive with each other. Now they stared at each other across a gaping difference in opinion. With the clock ticking. If the hydrocarbons had breached the casing, the gas would be rising this very moment through the formation. Someone had to decide—within hours—whether to reopen the well.

And here the Unified Command frayed. The government had the authority, but BP had the technology. BP wanted to keep the well shut in. The Chu advisers leaned toward reopening it.

Would the government really tell BP to resume spilling oil into the Gulf of Mexico? The thought horrified BP. For God's sake, the company had just stopped the damn gusher in a maneuver celebrated in the corporate boardroom as well as the Houston war room.

"Both parties were suffering from the uncertainty of who was in charge of deciding the next steps, and who would be blamed if things got bad," Hunter told me later. "BP quickly marshaled its forces, analyzed the data, and decided that the well was competent—a case they would defend at all cost. The government team was dispersed from Washington to Houston to Albuquerque to Menlo Park, communicating by telephone, and mostly erring on the conservative side by assuming the well must be reopened. In the midst of these two polar positions was the advancing clock."

Finally, Coast Guard Rear Admiral Kevin Cook, embedded in Houston, made a compelling argument. All along, the plan had been to keep the test going for twenty-four hours if the pressure was in the ambiguous range. The Unified Command had agreed to that protocol. Let's stick with the plan, Cook said.

Chu and his colleagues agreed that they wouldn't pull the plug immediately. But if they were going to keep the well closed for the full twenty-four hours, they would need sound scientific analysis that could support the hypothesis that the well was intact. They needed an independent verification of BP's scenario of reservoir depletion. BP wasn't a disinterested party here: the company would be fined per barrel spilled. The government needed someone with no skin in the game to say that BP's scenario was, at the very least, plausible.

It was time for emergency geological diagnostics, some tight-deadline heroics from the kind of people who, for whatever reason, devote their lives to studying the motion of fluids inside the earth. The government needed to get its best hydrologists and fluid dynamicists into the game.

The US Geological Survey has a lovely campus in Menlo Park, California, just a few miles from the San Andreas Fault, which ruptured catastrophically in 1906. The presence of that massive plate boundary, a strike-slip fault where the North American and Pacific tectonic plates fitfully slide past one another, gives some urgency to the generally serene research of the scientists on staff. Some of them had been brought into the Deepwater Horizon response as part of a well integrity analysis team. One of them, Paul Hsieh—pronounced "Shay"—now found himself in the thick of the conversation about what they were seeing with these ambiguous pressure readings.

Hsieh, fifty-six, is a hydrologist. He has worked on water issues between states and has published in such journals as the *Journal of Geophysical Research, Water Resources Research,* and *Geofluids.* Among his recent publications:

"Numerical Models of Caldera Deformation: Effects of Multi-Phase And Multi-Component Hydrothermal Fluid Flow."

"Evaluation of Longitudinal Dispersivity Estimates from Simulated Forced- and Natural-Gradient Tracer Tests in Heterogeneous Aquifers."

He'd been asked to join the science team back on June 20, when they were first thinking about well integrity in preparation for possibly switching caps. But as he sat in his office in Menlo Park, he lacked access to the pressure data that his colleagues were seeing in Houston. His colleagues needed his input, his analysis, but he needed to see the pressure curve, needed to see how the well had reacted to the shut-in. BP did not put this kind of highly sensitive data on the Internet. There was no simple, fast way for someone outside the BP building to obtain it.

A cell phone solved the problem. US Geological Survey scientist Steve Hickman, up on the eighteenth floor in the BP building, stood in front of a large LCD projection screen showing the pressure curve of the well. There were two lines showing pressure readings from two instruments. Hickman used his cell phone camera to snap an image of the screen. He then sent the cell phone image to Hsieh in Menlo Park.

It wasn't the standard way to conduct science, but it worked for what Hsieh needed to do. He had computer models. He added the pressure data to other known facts and assumptions about Macondo, including the flow rate, the oil viscosity, and the size of the reservoir.

Hsieh did not need anyone to tell him that the stakes were high.

"I had a sense that people were very nervous. This was the make-or-break point. I was nervous," he told me. "I would have liked to have had a week to do this. But I had eight hours."

He went on: "In all my other reports, you do your computer model, you come to a meeting, people talk about it, you have weeks or months to make revisions if somebody didn't like something, and even up to the publication date, you can make revisions and clean up things. But here, you do this, and people are going to make a decision. If you screw up, there's no recovery. That's the nerve-racking part of it."

He stayed late at work. His colleagues left. Soon it was just him and the janitor. Then the janitor left. Hsieh stayed deep into the night. He wasn't going anywhere. He lived near the office and could conceivably drive home for a quick nap, but no, he had to keep working on his model to find some way to show that maybe the reservoir really had depleted—and the oil spill could be over.

"I had this vision," he said.

It was a vision of Mars. No: a Mars probe. A NASA spacecraft. It was called the Mars Climate Orbiter. The spacecraft was a triumph of engineering, launched in December 1998 on a mission to study the red planet. On September 23, 1999, the spacecraft vanished as it prepared to go into orbit around Mars. Something was off in its navigation; it encountered Mars at too low of an altitude and burned up and disintegrated in the planet's upper atmosphere. The NASA engineers, heartbroken, received stunning news about what had gone wrong: In the communication between engineering contractors, there had been a failure to make a conversion between an English unit of measurement (pound-force) and a metric unit (newtons).

So that was on Paul Hsieh's mind. Mars. Crashed spacecraft. The oil industry uses English units, but Hsieh and the rest of the scientific community use metric.

Barrels become cubic meters.

Pounds per gallon become grams per cubic centimeter.

Pounds per square inch become pascals or kilopascals.

"I had to keep triple checking all these conversions."

He didn't need to drink any coffee: "Just the adrenaline of this responsibility was enough to keep me awake."

By the next morning, Hsieh had put together a PowerPoint presentation that, in effect, supported the plausibility of the BP hypothesis of reservoir depletion.

Hsieh's initial computer run presumed a square reservoir. In the model, the pressure leveled off after a few days. But Macondo's pressure continued to rise, albeit very slowly. Hsieh realized that he needed to revisit the model. His square reservoir had been accurate enough for the first hours after the shut-in, as he explained to me in an email:

> Suppose you throw a tennis ball and want to model the ball's flight path. During the first 0.1 second after the ball left your hand, the path can be simulated regardless of whether you are in a

> bedroom or in an auditorium. As long as the ball hasn't hit a wall, the paths are the same in both cases. However, after a few seconds, the paths would be different. In the bedroom, the ball might hit a wall and bounce off in another direction. In the auditorium, the ball would continue on its path.

Eventually, he tweaked the model to reflect a long, skinny reservoir. In a "channel" reservoir, it takes much longer, perhaps months, for the pressure to reach equilibrium in a shut-in situation. Moreover, that was the geological reality, as Hsieh learned from an expert in gulf geology, Peter Flemings of the University of Texas: The oil reservoirs were in places where sediments had been laid down millions of years earlier in underwater channels. The geological history of the gulf once again proved to be more than esoteric knowledge.

Friday morning, July 16, Hsieh made his presentation in a conference call with the scientists and engineers in Houston. BP's people made their own presentation arguing to keep the well closed.

Chu's team wasn't completely satisfied and still worried about the blowout scenario. But Chu would not call off the test quite yet. The well could remain closed—for now, but only if BP agreed to more monitoring, more staring at the seafloor for any hint of gas seeping from a dyspeptic, integrity-lacking Macondo.

Did they know anything for sure?

How did they know what they thought they knew?

Geology meets epistemology—always a recipe for sleepless nights.

The morning of July 16, President Obama cautioned reporters not to get too celebratory.

"As we all know, a new cap was fitted over the BP oil well earlier this week. This larger, more sophisticated cap was designed to give us greater control over the oil flow as we complete the relief wells that are necessary to stop the leak.

"I think it's important that we don't get ahead of ourselves here. One of the problems with having this camera down there is that when the oil stops gushing, everybody feels like we're done—and we're not."

Obama knew what the scientists knew, that this was still an exceedingly iffy situation. Thad Allen thought the well could be kept shut, but he also knew that the concerns of the science team had to be taken seriously.

"It was touch and go for the first eighteen hours," Allen told me. He positioned himself squarely between BP and the science team and tried to figure out how to get to yes. He was on the phone with, or texting, BP's Bob Dudley throughout the process. He'd listen to BP's arguments, and then say, "That's nice, but here's what's going to happen; here's the bar you have to jump over."

The science team agreed to keep the well shut in temporarily—with reopening still the default position—provided that BP did even more monitoring. There was some dismay on the government side at what seemed to be BP's foot-dragging on ROV monitoring of the seafloor.

Admiral Allen told Dudley that the scientists demanded more monitoring, more frequent seismic runs. But BP worried about collisions with so many ships right at the source. The company wanted to protect itself in case something terrible happened with all these ships operating so close together. So Dudley said to Allen, "Put it in writing."

Allen did so. Very late that Friday night, Allen sent Dudley an email:

> Dear Bob . . . I tried all of your numbers without success. I just finished a conference call with our principals. I need to spell out the expectation for operations tomorrow very clearly to you and ask that you give this your personal attention.
>
> The issue of monitoring for leakage and other indicators of well integrity has emerged as a litmus test for BPs commitment to moving the Well integrity test forward . . .
>
> You will have to commit to and execute a routine surveillance of the wellhead area until advised otherwise. That surveillance will consist of a double pass by a seismic array followed by a NOAA vessel capable of acoustic testing for methane. This will be conducted twice a day as a condition of further testing, and failure to adhere to these guidelines will result in an immediate order to cease the test and move to containment production.
>
> The exact details of the surveillance to be executed will be

provided in a follow on email reflecting Secretary Chu's personal direction.

This guidance is not discretionary and should be considered as a condition of continued operations.

Please call.

Thad.

Allen was telling Dudley, No messing around. You're a nanometer from having your well reopened. You have to prove that this isn't going to backfire. This is how it's going to be. *This guidance is not discretionary.*

Then came the anomalies. Bubbles from around the blowout preventer. Something puffing from the seafloor. A methane seep about three miles from the blowout preventer.

These things had the full attention of the scientists and engineers. The bubbles, BP said, came from nitrogen being released by cement around the casing as it cooled. With the well static, everything had cooled down dramatically, and the cement underneath the wellhead had contracted in a process that emanated bubbles of nitrogen. But the ROVs would have to sample some bubbles to make sure they weren't composed of methane, which would indicate a leak from Macondo.

Saturday afternoon brought the end of a forty-eight-hour period envisioned for the test. Would the well be reopened? No. But Thad Allen, in announcing his decision to extend the test, used ambiguous language that confused the news media. It sounded as though the test would have one more day and then the well would be reopened.

The BP folks sent a different signal. Kent Wells said, "Right now we do not have a target to return the well to flow." In a conference call with reporters, BP's Doug Suttles implied that the default position was to keep the well shut in: "We are just taking this day by day, and could be that we take it day by day all the way to the point we get the well killed." When Chu heard what Suttles said, he was livid. The default position was reopening the well.

During a conference call on Sunday, BP went down the list of negative indicators that so far had *not* presented themselves during the test. For instance, the pressure hadn't leveled off below 6,000 psi. (*Yeah,* thought Chu, *but the range of 6,000 to 7,500 was still ambiguous.*)

There was no visual or sonar evidence of broaching anywhere near the well. (Yeah, but the NOAA *Pisces* data covered less than 10 percent of the area around the well.)

There was no observed leak at the blowout preventer. (Yeah, but what about these bubbles, which might or might not be benign?)

There was no evidence from seismic surveys of something happening in the formation that resembled a blowout. (Yeah, but the seismic surveys were covering only about half the area within a mile of the well.)

"We are not in a totally good place, and there may be ominous signs," Chu wrote his colleagues. "We agreed to continue shut-in as long as we watch closely so that we can catch a leak early."

Anyone observing this process from a distance had to find it confusing. To complicate matters, the government released a letter that Thad Allen sent to Bob Dudley that Sunday, the eighteenth. What Allen wrote generated some alarming headlines:

"Given the current observations from the test, including the detected seep a distance from the well and undetermined anomalies at the wellhead, monitoring of the seabed is of paramount importance during the test period . . ."

A seep . . .

Undetermined anomalies . . .

It was more bad news. This integrity test looked to many people like a complete bust. No wonder that Monday morning, the nineteenth, a NOAA scientist received an email from the producer of a Canadian television news program:

"We would be interested in a five-minute live interview regarding the latest setback in the BP oil spill."

A *New York Times* science blogger opined, "Maybe it's time to bring in the Russians, whose Mir submersibles can go far deeper than the gulf well."

A writer for *Time* magazine declared that "this entire awkward series of command via letters and releases underscores how dysfunctional this response has been . . . Not exactly Apollo 13—and not exactly comforting."

Almost lost in the collective gloom was the news that the well remained sealed, and now, four days after the integrity test, no one was trying to reopen it. The well seemed to be passing the test, seep and anomaly notwithstanding. The spill no longer was all those scary things that Thad Allen had called it back on May 1: indeterminate and asymmetrical and anomalous. The crisis now had a parameter. It had a bracket. It wasn't over, but, pending some sudden reversal of fortune, the Macondo well would no longer spill oil into the gulf.

The expectations management strategy of the government and BP had been all too effective. The Sisyphean task of fighting the well had become so familiar to the public that the breakthrough moment was met with disbelief. The plume was gone—and indeed, it was never to be seen again—but optimism had not yet infiltrated the narrative of the nightmare well and the incompetent oil company and the ineffectual government. The Macondo rules still applied, for now, at least in the national conversation about this oil spill response, this festival of fecklessness. Success was not yet a plausible option.

Chapter 12

The Banality of Catastrophe

On July 19, even as the government and BP stared nervously at the seafloor and hoped that they could leave this damn well shut in, something a little bit like a smoking gun materialized in the investigation of the Deepwater Horizon blowout. Finally, we had a plausible explanation for one of the things that went wrong back on April 20.

On this Monday morning, at the Radisson Hotel just off Interstate-10 near Louis Armstrong New Orleans International Airport, the government investigators on the Marine Board—the joint probe by the Coast Guard and what was now known as the Bureau of Ocean Energy Management, Regulation and Enforcement (BOEMRE)—opened another set of hearings into the "casualty," as they called the April 20 tragedy.

The physical setting screamed government discount: The Radisson provided coffee, water, and afternoon cookies, but otherwise this was a no-frills environment. The lawyers sat shoulder to shoulder in multiple phalanxes, facing the board members at the front of the room. The testimony could be tedious, but it could also be revelatory. One had to pay attention closely lest one miss the telltale click of another tumbler turning in the lock.

The logistics of the event did not, unfortunately, include simultaneous translation of Engineer into English. Those of us from outside the petroleum industry struggled to understand not only the answers but also the questions. For example, Jason Mathews of BOEMRE questioned Lance Moore John, an employee of the contractor Weatherford:

> Q: Do you know what Weatherford equipment was actually used in this well?

A: Uh . . .

Q: Reamer shoe?

A: Reamer shoe, float equipment—

Q: Float collar?

A: Yes.

Q: Float collar was yours. How about the darts for the wiper plugs?

A: I think so.

Q: Who on the rig for Weatherford is actually responsible for the operation of those devices?

A: Nobody.

Q: Nobody?

A: We're not in charge of it. They just—we make it up because it came out there already pre-bucked. We just make it up and run it down the hole, and then they do their plug and cement job. I don't know exactly who does the plugs or what.

Q: You obviously know what the term "float collar" is, correct?

A: Correct.

Q: In the float collar, what pressure should it take to shear the pins to convert it from a fill-in device to a check?

A: I have no idea.

Q: No idea. Do you know what pressure it did shear out at?

A: I have no idea.

And so on, like that: simple questions, simple answers, but roughly as comprehensible to the lay person as *Finnegans Wake*.

And yet the miracle of the Marine Board hearings was that, over time, revelations burbled to the surface. This is where we met the players in the tragedy: the managers, engineers, mechanics, drillers, navigators. The bosses and the bossed. The captain of the rig, the VIPs, the guys working the pencils back at the home office.

And on this Monday afternoon, we heard from a mud engineer, Leo Lindner. Lindner was a technically a "drilling fluid specialist," and worked on the Horizon for the contractor M-I Swaco. Questioned by investigator Mathews, Lindner recounted the uncertainty around the negative pressure test.

Recall that, from the early days of the disaster, the inquiry into what went wrong centered on this "negative test," the key diagnostic test of the quality of the cement job. The initial negative test on the afternoon of April 20 produced an unwelcome result: excess pressure up the drill pipe. When the crew bled off the pressure and closed in the pipe, the pressure rose back to 1,400 psi. There should have been *no* excess pressure, because the well had been cemented. The BP company men decided to redo the test, this time on the kill line, and they got a much more favorable result. The well flowed briefly and then stopped, with no flow observed for thirty minutes. They then proceeded with the mud displacement. Heavy mud out, lighter seawater in. And the gas came up the well.

Lindner said he knew something wasn't quite right with the initial negative test, and he ordered the crew to stop off-loading the mud from the well in case they needed to "reverse" it and put the mud back down the riser. "I was kind of hedging my bets," he said. Then the former college English instructor discussed the "spacer" he mixed up for use during the mud displacement. *Spacer* is the generic term for a thick fluid that's inserted into the well to keep the mud and seawater separate. Lindner testified that the displacement procedure was an unusual one. A normal spacer is about 180 to 200 barrels, he said, but, with BP approval, he combined batches of two different fluids to create a spacer that was close to 450 barrels—a double "pill" of spacer, as he put it. The two fluids were known as Form-A-Set and Form-A-Squeeze, and are known in industry terminology as lost circulation material (LCM). These heavy fluids are designed to firm up holes and reduce mud losses of the kind that had plagued the Horizon during the Macondo job.

Q: Why was there such a large spacer?

A: The day I got to the rig, I made the two LCM pills, and Mickey had wanted to use those instead of trying to build another spacer. When we combined the two fluids, we had a very large spacer.

> That was vetted through BP's environmental people. The drilling engineers were involved and the—our fluid people were involved; their fluid people were involved. So that's why we used so much . . .
>
> Q: Is it common to have a lost circulation material as a spacer?
>
> A: Not common.
>
> Q: How common is it?
>
> A: It's an idea we—I say, we, but the rig had kicked around if we were in that position . . .
>
> Q: When running a spacer with that type of—I mean, that type of spacer, do you have any issues with your choke and kill line[s] with any type of possible clogging or plugging of the choke and kill lines?
>
> A: I didn't foresee any issues, no.

It was mentioned obliquely, but here was a solution to the riddle, potentially, of why the negative test had been misinterpreted. The crew, with BP's approval, had sent 450 barrels of a peculiar concoction of dense fluid down the well, using it as the "spacer" that keeps the valuable mud from being contaminated with seawater. This 16-pound-per-gallon gunk was supposed to stay above the blowout preventer, but it could have sunk through the much lighter seawater, gotten into the blowout preventer's kill line, and clogged it. That could explain the lack of flow on the kill line during the second negative test. The Horizon crew members thought they were seeing evidence of a good cement job, and this test result led them to ignore the anomalous 1,400 psi that remained on the drill pipe.

Why use this particular recipe of sludge in the well? The answer: to save a little time and money, thanks to a loophole in the environmental laws. This came out as Lindner was questioned by David Dykes, the co-chair, also from BOEMRE:

> Q: You had a 450-barrel lost circulation material pill that you had combined out of two separate pills. So take an average, if you have two, you have two 225-barrel pills, originally?

A: I don't remember exactly what the pit volumes were, but yes.

Q: Somewhere in that ballpark?

A: Somewhere in that ballpark.

Q: Why didn't you only use one and just dump the other one overboard?

A: We couldn't dump overboard because of the way BP—well, and this is, I'm speaking for BP. This is my understanding of it. That according to the RCRA Act, they couldn't just discharge something that hadn't been circulated through the well, the wellhead—

Q: The well bore?

A: —because it wouldn't be an exempt waste.

Q: So to be in compliance with the National Pollution Discharge Elimination System requirements, also known as the NPDES?

A: I don't know that, plus I am not an environmental lawyer, but I thought the RCRA Act was different than the NPDES. But it is an environmental act, from what I understand.

Q: So the fluid has to come in contact with the well bore before it can be discharged overboard.

A: In order to be considered an exempted waste, right.

Lindner was saying that under the Resource Conservation and Recovery Act, enacted in 1976, Form-A-Set and Form-A-Squeeze are considered hazardous wastes. Normally, the fluids would have to be off-loaded to a boat for disposal on shore in a hazardous waste facility. But there was a loophole: Any water-based liquid that is circulated in the well bore can be dumped overboard. Dykes made sure he was hearing correctly:

Q: . . . to meet the requirements for an exempt waste, or exempt discharge, it's got to come in contact with the well bore. It's got to be used.

A: Circulated through the well, yes.

The final questions came from Greg Linsin, the attorney for the flag state, the Marshall Islands.

> Q: Is that standard practice to combine those two different types of spacers?
>
> A: It's not something we've ever done before.
>
> Q: Do you know what the chemical consequences are of combining those two spacers?
>
> A: I ran a pilot test on both spacers, mixing them, and ran the rheology. It didn't set up. I kept the sample overnight to see if it would do anything, and it didn't. Other than as opposed to something done in our office or technical people, I'm not aware of that.
>
> Q: What was the volume of the sample that you ran?
>
> A: A gallon of each, mixing the two.
>
> Q: And based on the mixing of a gallon of each, you didn't see any consequences; is that correct?
>
> A: I didn't see anything—I didn't see it setting up or anything.
>
> Q: It didn't set up?
>
> A: Right.

And with that final revelation—that he had mixed a gallon of this and a gallon of that the night before the blowout, and waited to see if it would harden, or "set up"—Lindner was dismissed.

During the break, Ronnie Penton, the attorney for rig worker Mike Williams, told me and my colleague David Hilzenrath that the conclusion was obvious: "That large pill skewed the testing." The spacer threw off the negative test.

This was, indeed, a potential answer to the negative test mystery. BP's internal investigation would eventually come to a similar conclusion. But there were other possibilities. Perhaps, during the second negative test, someone had inadvertently closed a valve that cut the flow to the kill line. Or perhaps gas from the well flowed into the kill line, which was very cold due to the deep water, and formed methane hydrates that clogged the kill line.

I asked Richard Sears, the deepwater oil industry veteran working with the presidential commission, if he thought the unusual spacer fluid was a key factor in the misinterpreted test and, ultimately, the blowout.

He shrugged.

"It just added another little risk element into the thing. You had these unusual fluids now. You had this column in the well that was all these different types of fluids stacked on one another," Sears said.

Was the improvised spacer a little bit cockamamie?

"It wasn't cockamamie. It was ad hoc," Sears said, then added: "What's the difference between cockamamie and ad hoc? You tell me."

In the view of Sears, risk builds like plaque. It's cumulative. There's a fundamental mistake that people make: They think that, as they deal with one risky situation after another, they can look at each decision in isolation. That's what happened with the Horizon, he said: The individual decisions were generally reasonable. But the risk was accumulating. And the people who had to interpret the negative test were not necessarily in the loop on all the previous decisions and judgments. They didn't know what we know now about the matrix of decisions across time and space, decisions that ranged from the rig to Houston, and from BP to Transocean to Halliburton to M-I Swaco and other contractors.

Even if the spacer is a red herring in the Macondo mystery, it provides a reminder that there rarely is such a thing as a job that is completely "by the book." The Macondo job was a unique adventure. Remember Interior Secretary Salazar saying that there had been forty thousand wells drilled in the Gulf of Mexico? Here's what Richard Sears says about that:

"Show me the thousands of wells that look just like this one, and were drilled and completed and temporarily abandoned just like this one."

What had Lindner said?

It's not something we've ever done before.

Not My Job

If it's hard to find the incontrovertible smoking gun in the Deepwater Horizon disaster, that's because there's smoke coming from so many places. It's hard to know which factors were integral to the disaster and which were irrelevant, but a number of things grab our attention:

- a minimalist well design
- questionable foamed cement
- concern about schedule and cost
- a blowout preventer with maintenance issues
- an unusually deep "surface" cement plug
- only six centralizers despite warnings of a *severe* gas flow potential
- simultaneous operations that complicated efforts to monitor the fluid volumes going in and out of the well
- no cement bond log test for flaws in the cement job
- significant changes in the well plan in the days before the blowout
- changes in key personnel on the rig

And as the Marine Board continued its hearings over the summer and into the fall, more anomalies surfaced, more problems that may have steadily and stealthily guided the rig toward its doom.

We would eventually hear from Jesse Gagliano, the Halliburton employee embedded with BP's well team in Houston, and the author of the report that warned of the *severe* gas flow. Gagliano never imagined that the well would blow out. The "severe" warning was only an alert for possible channeling, which would require remedial action.

Those of us trying to make sense of events have to remember that history looks backward but life is lived forward, in the thin blade of the "present," with a limited attention spotlight, and countless distractions, the daily routine easily compromised by lack of sleep or hunger or emotional distress, and we are not always logical and diligent. This is not to excuse the people whose actions may have contributed to the blowout, but to recognize how error can be a natural part of any human enterprise. No malign intent is necessary.

Wouldn't it be swell if all our decisions were good ones, made with abundant information amid unambiguous circumstances? The completion and abandonment of the well seemed safe enough for veteran rig workers to be in the shower, in bed, reading a book, having a smoke, chatting on the phone, sleeping. The world as they knew it was coming to an abrupt end, but somehow they missed the signal, they were looking the wrong way, or they simply refused to believe that what they were seeing was real. *She just blew,* Jimmy Harrell had said that awful night as he watched his beloved rig burn on the serene waters of the Gulf of Mexico.

Back to the negative test: Why did the drill team and BP well site leaders on the afternoon of April 20 not recognize that the 1,400 psi on the drill pipe was evidence of a failed cement job and a flowing well? Investigators for the oil spill commission later determined that the 1,400 psi on the drill pipe was exactly the pressure that would be exerted by the roughly 12,000-psi Macondo reservoir given the amount of mud and seawater filling the 18,000-foot well. The 1,400 psi wasn't a hint of something wrong, it was the mathematical proof of it.

But the people on the drill floor didn't know that. They operated with limited information. People on the rig didn't know everything that people at the BP office in Houston knew; people at the BP office in Houston didn't know everything that people on the rig knew. Did the drill team know, as it studied that anomalous 1,400 psi pressure on the drill pipe, that a Halliburton model had warned that, with so few centralizers used in the cement job, the well had the potential for a *severe* gas flow problem? What would their analysis have been in that case?

The critical decisions regarding the final pressure tests on the afternoon and evening of April 20 remain extremely controversial and the subject of litigation. It is not clear to what extent BP personnel in Houston monitored the situation in the hours before the blowout. Shortly before 9:00 p.m., Don Vidrine had a discussion by phone with Mark Hafle, an engineer on the Houston-based Macondo well team, and told Hafle that the pressure test issues had been resolved, according to the chief counsel's report of the presidential oil spill commission. The report concluded: "Given the risk factors attending the bottomhole cement [BP's cement job], individuals on the rig should have been particularly attentive to anomalous pressure readings. Instead, it appears they began with the assumption that the cement job had been successful and kept running tests and proposing explanations until they convinced themselves that their assumption was correct."

One of the people who were there that day, in the drill shack, listening to the conversations, was Lee Lambert. Lambert appeared before the Marine Board on July 20, the day after Lindner's revealing spacer-fluid testimony. Lambert was on the rig in a training capacity when the well exploded. His title, he said, was well site leader of the future, Deepwater. Under questioning, he said he was aware of the cost to BP of drilling the Macondo well: "I understood at the time that the overall spread rate was close to one million dollars a day."

But, even though he was there to learn, to listen, and to watch, by his own testimony he was oblivious to a long list of anomalies compiled by the investigators. Mathews, reading from a document prepared by investigators, tossed one anomaly after another at Lambert to see if any of them registered:

> Q: I'm going to go through a list of anomalies . . . "Had the pressure up on the cement shoe valves nine times to convert from the filling line casing to act as a check valve, per the BP morning report."
>
> A: I'm not aware of that.
>
> Q: Okay. On April 19 at 5:30 to 19:30, it says: "Low pump pressures. Return flow after float shear out appeared to be high."
>
> A: I'm not aware of that.
>
> Q: On April 20, beginning at 0300 hours to 0400 hours: "Running string for the 9 and 7/8ths by 7-inch string casing pulled wet after the seal assembly test was completed. Running string pulled three strands wet and they—so they slugged the well bore with 30 barrels and a 16.3 pound per gallon pill."
>
> A: I don't specifically recall that.
>
> Q: Okay. Do you recall on April 20 at 1200 hours a "15 barrel gain in a trip tank after the casing test pressure was released, this

> seemed a little large for pressure bleed back for fluid compression and would expect only a 5 barrel gain"?
>
> A: Not aware.

Lambert, by his own testimony, saw nothing, heard nothing, knew nothing that could be considered a sign of serious problem, much less a calamity.

He was privy to a key conversation about the 1,400 psi pressure on the drill pipe during the first negative test. He said the toolpusher, Jason Anderson, offered a possible explanation for the pressure, something that Anderson called the bladder effect, or annular compression.

> Q: And what was Mr. Anderson trying to describe the bladder effect to be?
>
> A: That the heavier mud in the riser would push against the annular and transmit pressure into the well bore, which in turn you would see up the drill pipe.
>
> Q: Did you take the bladder effect, or the annular compression, to be a valid description of what was going on at the time?
>
> A: From my position, I took it as a valid explanation. This is the first time I've seen a negative test, so.
>
> Q: It was the first time you had ever heard of bladder effect?
>
> A: Yes.

But Jason Anderson could not give his version of the conversation, could not answer questions, could not explain what he meant by this alleged comment. He was among the dead, killed in the line of duty as he tried desperately to shut in the well. So was the driller, Dewey Revette, who was also right there in the drill shack during the conversation.

The BP well site leaders, Kaluza and Vidrine, would not testify. As the company men, they had the final authority to interpret the negative test. Kaluza apparently embraced the "bladder effect" scenario. "I believe there is a bladder effect on the mud below an annular preventer as we discussed," Kaluza wrote two colleagues within a week after the explosion, and then elaborated on how it could have explained the anomalous pressure on the drill pipe, according to the chief counsel's report to the presidential oil spill commission. Someone forwarded Kaluza's email to Pat O'Bryan, the BP vice president of drilling and completions, who had been one of the four VIPs on the rig April 20. O'Bryan wrote back: "??? ???????????????????????????? . . . ."

His incredulity reflected what the oil spill commission investigators heard repeatedly from veteran drilling engineers: Bladder effect? There's *no such thing* as a bladder effect.

What there was, according to the chief counsel's report—a 371-page document released February 17, 2011—was a dramatic failure of management, communication, and training within BP and Transocean.

To take one example cited by the report: a Transocean rig had a near disaster on December 23, 2009, in the North Sea that had eerie parallels with the Macondo calamity months later. In the December event, huge quantities of mud and hydrocarbons shot out of the well and into the sea before the crew managed to seal the well and stave off an explosion. The crew had been lulled into complacency by a successful negative test and was displacing mud with seawater in a process that hampered the monitoring of fluids. The workers saw signs of a gas kick but discounted them. After the incident, Transocean developed new guidelines in which it warned drill teams not to let down their guard after what appeared to be a successful negative test. But the chief counsel's report concluded that Transocean had not adequately distributed the new warnings. There was no evidence that anyone on the Horizon had ever seen those guidelines.

Again and again, people didn't speak up, didn't talk to one another, didn't pass along warnings. The chief counsel's report states: "BP's onshore team should have, and easily could have, alerted the well site leaders and rig crew that cement failure at Macondo might be more likely than normal and instructed them to be extra vigilant regarding any odd pressure readings."

But here's the reality: Some people barely knew each other. Kaluza, for example, had been on the rig all of four days. There was sometimes confusion over who had authority to do what. BP had restructured its engineering teams, people were moving around, new people showing up, and all the while, the drilling plans kept changing. The situation exasperated Brian Morel, the young BP engineer assigned to Macondo. That was the message delivered April 17—three days before the blowout—by Macondo well team leader John Guide in an email to his boss, David Sims:

"David, over the past four days there has been so many last minute changes to the operation that the WSL's [company men] have finally come to their wits end. The quote is 'flying by the seat of our pants.' Moreover, we have made a special boat or helicopter run every day. Everybody wants to do the right thing, but, this huge level of paranoia from engineering leadership is driving chaos. . . . Brian has called me numerous times to make sense of all the insanity. . . .The operation is not going to succeed if we continue in this manner."

Sims did not react "with alarm or stop work on the rig," the chief counsel's report notes. Instead, he wrote back with a stay-the-course message:

"John, I've got to go to dance practice in a few minutes. Let's talk this afternoon. For now, and until this well is over, we have to try to remain positive and remember what you said below—everybody wants to do the right thing. The WSLs will take their cue from you. If you tell them to hang in there and we appreciate them working through this with us (12 hours a day for 14 days)—they will. It should be obvious to all that we could not plan ahead for the well conditions we're seeing, so we have to accept some level of last minute changes."

Sims added at the end of his email: "We're dancing to the Village People!"

One BP well site leader who had worked for many years on the Horizon, Ronald Sepulvado, appeared before the Marine Board and testified that he was not on the rig on April 20. He had left four days earlier, overlapping with his replacement, Kaluza, by just thirty minutes. He testified that after he flew to shore, he didn't check his email and turned off his cell phone. He saw nothing, heard nothing, knew nothing about the events of those four days leading up to the catastrophe.

The reason he left the rig, he said, was to attend a "BOP school" in Lafayette, Louisiana. He had to go to classes to learn about the prevention of blowouts.

Kill, Kill, Kill

The endgame was oddly muted. The sealing of the well on July 15 did not kill Macondo, except as a round-the-clock news story. The oil spill was still big news, to be sure, but just as the coming of the plume had created an extra layer of psychic angst on top of what was already very rationally understood to be a crisis, the disappearance of the plume eased the teeth-grinding anxiety of the situation. This became, if there is such a thing, a more tolerable catastrophe.

But it wasn't over, by any means. The well remained pressurized, unplugged, highly dangerous—still a hole threatening to explode. It was a couple of valve failures away from gushing anew. (Indeed, in early August a valve did fail, and only a little bit of friction in a backup valve kept it from reopening, according to *The New York Times.*) Meanwhile, the seismic and acoustic surveys continued to search for signs of something funky going on that could signal a leak into the rock formations. Someone at BP was assigned to count the bubbles of gas coming from tiny leaks in the blowout preventer and capping stack. ROVs focused on the sea floor as if in a stare-down contest.

A tropical storm, Bonnie, blew through the gulf and bollixed up everything at the source for a week or so. The Development Driller III had to halt the drilling of the first relief well and sail away before it could get the final string of casing cemented. Bonnie turned out to be no big deal, and many of the vessels never even left the disaster site, but the big rigs move glacially, and the Bonnie incident showed again how vulnerable the whole operation was to disruptions from tempests.

The relief well, as Thad Allen never failed to remind everyone, was the ultimate solution to the crisis. But BP surprised everyone with a new idea: Even before the relief well intercepted Macondo, BP would try to kill the well from the top again in what amounted to a do-over of the top kill. This would be a "static kill," because the well would not be flowing. In the new procedure, the mud wouldn't have to be pumped furiously into the well but could mosey into the well at a leisurely pace. After an initial rise in pressure, the well would see pressures fall as the mud pushed—or "bullheaded"—oil down into the reservoir.

Chu and some of the other scientists had little enthusiasm for the new plan. Chu felt jerked around by BP. The company had surprised the government with this static kill idea. Chu bristled at what he felt was a chronic lack of transparency on the part of BP. For example, according to Chu, BP had said during one conference call that the static kill would require raising the pressure by a "minuscule" (Chu's word) amount, but days later, the company said it actually needed 1,000 psi of leeway—a big difference.

"Sometimes they seem to change 'facts' to suit their purposes . . . not the kind of behavior that builds trust," Chu groused in an email to Garwin.

Garwin was adamant that BP shouldn't attempt the static kill. He thought BP should stick with the relief-well operation as the first line of attack. The static kill, Garwin said, could raise pressures in the capping stack and the other hardware on top of the well to dangerous levels. Garwin did not support BP's latest hypothesis for why the top kill had failed. BP said the top kill mud simply shot out the top of the well and settled onto the sea floor. Garwin thought that unlikely, believing that mud was lost down in the reservoir. In his and the government team's scenario, the hydrocarbons were coming up the annulus, not up through the production casing (the long, narrow pipe running the length of the well).

Gizmologist Alex Slocum, echoing Garwin, argued that the smart path forward would be the production of oil from the reservoir. Rather than trying to plug the Macondo well, why not drain the reservoir in a controlled fashion? Maybe the rig that had been drilling the backup relief well (the Development Driller II, the rig I had visited in late June) could provide some real "relief" to the reservoir by drilling directly into it and siphoning the hydrocarbons to the surface. That would be, Slocum thought, a safer long-term solution. As he wrote to his colleagues:

"We may 'kill the well,' have a Miller Beer, and be so happy . . . and in a few years POOF, Macondo pains due to the fact we reallllly do not know whats up with that beasty bore . . ."

Some of the full-time government scientists in Houston, including Marcia McNutt, were more bullish on the static kill. It would help diagnose the condition of the well and make the bottom kill much easier, she argued.

There was another factor on BP's mind: The engineering protocol for the relief-well operation required that the Macondo well be reopened briefly at the top—flowing anew into the gulf—just prior to the bottom kill. There were sound engineering reasons to vent oil from the stack, and it wouldn't be much oil compared to what had leaked already. But the plume would be back. The plume! The hideous oil geyser from the depths of hell! It would surely incite an epic foofaraw in the news media.

So this was the integrity test debate all over again: The science team generally on one side of the argument, BP on the other.

The person trying to get everyone to agree on a path forward was, as ever, Tom Hunter. Hunter—the unsung hero of the crisis, in Thad Allen's opinion—sought to establish areas of agreement so that the conflicts between the government and BP could be handled piecemeal. In Hunter's view, this was no time to let resentments about the process foil what might be a good strategic move. His philosophy: Stick with the facts, stick with the science, don't get caught up in assumptions about the motivations of others. "Despite biases, different levels of understanding, and disparate opinions, it is possible to forge a reasonable and in this case proper path forward," he told me.

Chu finally decided that the static kill was worth the risk. The operation was a go.

On August 2 the Development Driller III rig finished cementing the casing on the relief well, enabling the rig to withdraw the drill pipe and sit tight while the static kill commenced. On August 3, the 106th day of the crisis, BP performed an "injection test," putting a fluid known as base oil into the well from the top. The pressure increased only slightly, then began to drop—just as the engineers had hoped. BP quickly followed with the heavy drilling mud. The pressure continued to drop. At eleven o'clock that night, Macondo was choked with mud. On the night of August 4, BP followed the successful mud shot with a massive dose of cement.

Macondo now had a mile of cement in its gullet.

The Obama administration was not about to send BP a congratulatory bouquet. The good news put the administration in the awkward situation of needing to announce the positive development without suggesting that BP had done something right. The White House pushed the notion that Chu and the government team had accelerated a sluggish response. From the way the administration sometimes discussed its role, one might suppose that Chu had personally plugged Macondo with his bare, brilliantly nimble hands.

On August 4, with Macondo already choked with mud, administration officials briefed reporters on an instantly controversial government report called *Oil Budget: What Happened to the Oil?* The government scientists had already concluded that the well had initially flowed at 62,000 barrels a day and, as the reservoir depleted, dropped to 53,000 barrels a day at the time of the shut-in. If the government was correct, Macondo had spewed 4.9 million barrels of oil, with 4.1 million barrels flowing into the gulf and the rest captured by BP with its top hat and other containment devices. BP never accepted these numbers and would come to argue that the flow might have been only half that estimated by the government. In any case, here in early August the pressing question was: Where did all that oil go?

"The vast majority of the oil from the BP oil spill has either evaporated or been burned, skimmed, recovered from the wellhead or dispersed, much of which is in the process of being degraded," the National Oceanic and Atmospheric Administration announced in releasing the report. The government's initial assessment showed that an astonishing amount, 25 percent, had naturally evaporated or dissolved, and another 16 percent had been naturally dispersed. BP had recovered 17 percent directly through containment operations, another 8 percent had been chemically dispersed, 5 percent burned, and 3 percent skimmed. That left only 26 percent of the oil unaccounted for. A raging debate erupted over whether the oil was truly "gone" (cue debate over whether oil that has been "dispersed" is there or not there, gone or not gone) and the fact that the *Oil Budget* had not been peer reviewed by scientists in any traditional sense.

The central point of the Oil Budget was rather startling given what everyone had witnessed over the previous three and a half months: This was not going to be the end of the Gulf of Mexico. The gulf had not been painted black.

A sad footnote to all this is the tale of Matt Simmons. The famed energy investor had been perhaps the most prominent doomsayer during the spill. On August 8, 2010, Simmons died in his vacation home in Maine. He drowned in his bathtub. The medical examiner's office ruled the death accidental, with heart disease as a contributing factor. Simmons was sixty-seven years old.

During the August 4 press conference at the White House, a member of the media put a question to Robert Gibbs:

"Robert, back in May, Tony Hayward said, quote, 'The Gulf of Mexico is a very big ocean. The amount of volume of oil and dispersal we are putting into it is tiny in relation to the total water volume.' After he said that, the president said that he would have fired Mr. Hayward if he said those comments. Now it now appears that Mr. Hayward may, in fact, have been right. Does the administration owe him an apology?"

Gibbs shot back "No. I don't think he was right. I mean, let's understand that a third of—a third of what is captured was—was based on—directly on a containment strategy that had to be constructed and—I'll say this—a containment strategy that—that we pushed BP forward on, that we pushed BP to accelerate in order to capture the oil that was leaking.

"Nobody owes Tony Hayward an apology . . . The apology that is owed by—any apology that is owed is to the disruption to the lives of families, fishermen, hotel owners, people that grew up in and—and—and understand the beauty that is the Gulf of Mexico. That's the apology."

Next question: "How much of this success of today that you're laying out for us is attributable to BP and how much to the federal government? Do you think that you had to push BP to get here? I see you're nodding your head."

"Well," said the press secretary, "again—again, I think if you go back and look at the directives signed by Admiral Allen to various people in the corporate structure at BP, we asked for and demanded that particularly their containment strategy be accelerated. We asked for and demanded that not one relief well be drilled, but two, in order to ensure an amount of redundancy in the system that would allow for a mistake or an error.

"I think that the response as it is would have been different had Admiral Allen and others—Carol, Jane, the scientific team, Secretary Chu—not pushed at every step of the way for BP to do—to do things more comprehensively and faster."

That was the administration's take on it. But from the BP standpoint, the government had been a speed bump, not an accelerant. I asked a BP spokesman if the government's flow rate work spurred the company's deployment of the 3-ram capping stack as part of the expanded containment. The official corporate answer: "From the early days of the response, we pursued multiple parallel paths to contain the oil (as much as possible), and our efforts did not change based upon the government's focus on flow rate."

BP has remained cagey, its narrative heavily lawyered. The comprehensive story of the Deepwater Horizon disaster and the subsea response will not be written until the people in the black box emerge into sunshine and tell their story without regard to corporate liability or stock price.

The government did not plug the well, even though it contributed at key moments. This was, ultimately, a BP-created problem, and it had a BP-created solution. Which is what the oil spill law demanded.

Even if the government didn't plug the hole, however, the government did its job, and arguably did it with a great deal of aplomb given its weak opening position. The scientists, led by Steve Chu and Tom Hunter, performed a necessary role of quality control. The government held BP's feet to the fire—a fire of its own making. The great Salazar line, "So our job is basically to keep the boot on the neck of British Petroleum," sounded like the silly bluster of a blowhard at the beginning of May, but over time it became a more accurate description of the relationship between the government and BP. The collaboration, tense and snippety as it was in public, worked well on the front lines, where professionals put their egos and any political considerations aside and simply did their jobs.

In a functioning society, this is what happens. The pros do what they have to do. They stay at the office late and get there early the next morning, unless the battle rhythm requires that they pull an all-nighter. One of the little-known secrets of the BP gulf oil spill and the response to the gusher is that the people involved were not actually incompetent, stupid, or evil. They were smart people who worked hard and did their best in an extraordinary situation.

Grown-ups were in charge. They worked together. The government and BP would never have a straightforward partnership, but what they were doing was something very much like the collaboration envisioned in the National Contingency Plan.

"Nobody understood it, and they didn't like what they saw," Allen said of the plan. But he has a verdict: "It works. It worked this time. It was harder because of the folks who didn't understand it."

The process, he realized, had been politically battered.

"Social and political nullification of the National Contingency Plan. That's what I call it," Allen said.

The public rancor was part of the special poison of the oil spill. Victory did little to diminish the pervasive bitterness of the story. The spill was offensive. Umbrage was taken. This could never be an Apollo 13 story, after all, never a glorious triumph over daunting odds, because the insult had been too deep and too ugly. There would be no forgiveness. It was just one of the unfortunate aspects of Macondo that the killing of the well would always be a thankless task.

But the people who were there, the people who worked all day and night to solve the problem and sweated out the anxious moments and had the guts to make the tough calls, don't need anyone to tell them what really happened. They fully own their personal history.

In a Harvard commencement speech in June 2009, Steve Chu offered graduates some advice:

"In your future life, cultivate a generous spirit. In all negotiations, don't bargain for the last, little advantage. Leave the change on the table. In your collaborations, always remember that 'credit' is not a conserved quantity. In a successful collaboration, everybody gets ninety percent of the credit."

Even after the static kill of early August, the well was not officially dead. There was much work to be done on the relief well, and as it happened, the government and BP found new chances to grind gears a little and look at one another askance. For the people charged with making sure the well was "dead, dead, dead," as Chu told me one day by phone, the summer of the spill was not yet over. But in many ways, the pressure was off. The story faded from the front page, just as Marcia McNutt had predicted it would. The relief well finally achieved its target in the third week of September. Reaching a depth of 17,969 feet, the drill team reported signs that it had intercepted Macondo—twelve barrels of mud lost, changes in the torque on the downhole motor.

On September 18 the rig cemented the annulus of the Macondo well, and the next day, Admiral Allen put out a statement: "[W]e can finally announce that the Macondo 252 well is effectively dead."

The old blowout preventer had already been taken away for testing. A new blowout preventer had been put in its place, but eventually that was removed, and the Macondo well was put through a formal process of plugging and abandonment, supervised by BOEMRE.

On November 8 there was nothing left of Macondo but the stub of a pipe sticking up from the wellhead. An ROV carefully placed a cap on the pipe stub, while another ROV filmed the operation, making sure to capture the words stenciled on the top of the cap—"April 20, 2010"—and the words on the side:

"In Memory of the Deepwater Horizon 11."

The Banality of Catastrophe

The death of Macondo did not end the story. There were investigations, reports, hearings. There would be Lessons Learned. And, of course, there would be litigation, expanding by the day, flourishing, with a life of its own, like something foaming from a Petri dish.

The season of reports began on September 8, when BP unveiled the so-called Bly Report, named after lead investigator Mark Bly. The Bly Report summarized the Deepwater Horizon tragedy as the result of eight distinct breaches of physical and operational barriers. Any one of those eight barriers could have prevented the calamity, Bly's team concluded. In this scenario, the disaster was a long-shot event, an improbable jackpot:

"[A] complex and interlinked series of mechanical failures, human judgments, engineering design, operational implementation, and team interfaces came together to allow the initiation and escalation of the accident."

According to BP's investigation, most of the failures and errors could be attributed to the contractors, with Transocean fingered in particular. The Bly Report concluded that the cement job by Halliburton failed because the nitrogen in the foamed mixture separated away in the hot environment of the deep well. Rather than having cement of uniform density, the cement became like the old three-in-one Jell-O, with density varying by layers. BP believes that some of the nitrogen may have infiltrated, and compromised, the nonfoamed cement near the bottom of the casing in what is known as the shoe track.

In the BP scenario, gas entered the annulus, flowed down to the bottom of the well, back up the inside of the production casing, through the shoe track, and on up past the blowout preventer and into the riser. As it rose, the gas expanded. BP faulted the rig crew for not recognizing signs of gas in the riser more quickly. When it became apparent that a well control event was underway, the crew channeled the hydrocarbons through a mud-gas separator near the top of the rig rather than through a pipe that could have diverted the explosive material overboard, the report stated.

The blowout triggered cascading failures. For example, the explosion damaged the electrical lines that connected the rig to the blowout preventer. When Chris Pleasant hit the Emergency Disconnect System (EDS), an electrical signal should have disconnected the riser from the well, which would have saved the rig by cutting the source of the hydrocarbons fueling the fire. Without power, the EDS was useless.

The Bly Report landed with a public relations thud. Ostensibly this was a purely technical document produced by fifty-plus investigators, many brought in from outside companies on contract. The team operated behind locked doors; no stray executive, supposedly, would be able to infiltrate the workspace and influence the findings. But Mark Bly also regularly briefed the top BP executives and the corporate board. The company lawyers were allowed to review the report before it went public, according to *The Wall Street Journal.* If Bly had any hopes that the review would be seen as objective, they were dashed as soon as the report hit the web at seven in the morning, eastern daylight time, and the media began pointing out that the BP investigators had determined that the blowout was largely someone else's fault.

Normal Accidents: Living with High-Risk Technologies is the title of a seminal 1984 book by Charles Perrow that explores the history of technological disasters and anticipates the Deepwater Horizon disaster to an eerie degree. Perrow talks about "normal" or "system" accidents in which the calamity emerges from the basic characteristics of a complex, interlinked system. In these accidents, backup safety elements are often not truly separate from the primary element—one mishap can take them both out. Perrow writes, "We have produced designs so complicated that we cannot anticipate all the possible interactions of the inevitable failures; we add safety devices"—think, blowout preventers—"that are deceived or avoided or defeated by hidden paths in the system."

Perrow argues that such accidents, though rare, emerge naturally from the characteristics of the technological system. The key feature of these accidents is that different components of the system are more closely linked than anyone can realize. They are "tightly coupled." Thus one failure causes another failure. You could think of it as dominoes falling, except it's more brutal than that: The dominoes are all simultaneously crushed from above. (To take an example that has crossed my mind a few times: A writer trying to put together a manuscript might decide to back up the file on his laptop with a full set of up-to-date printouts, and then, to add yet another layer of redundancy, copy the electronic file onto a flash drive. But if he keeps the printouts and the flash drive in the same shoulder bag as his laptop, he doesn't truly have a doubly backed-up system, because a thief could swipe the whole thing in a single stroke.)

The Bly Report turned up numerous instances of hidden pathways and unseen linkages in the safety systems in deepwater drilling. Like: There are six emergency methods for activating the blind shear ram on the blowout preventer, including the dead-man switch that's supposed to work when the BOP loses communication with the rig. But that's not really six independent systems, because they all depend on the same ram. And if the ram is dysfunctional at some basic level, it doesn't matter if there are one thousand different methods for signaling it to close. BP contended that the all-important yellow and blue control pods on the BOP—the brains of the operation—were in bad shape even before the blowout. BP concluded that "there was a fault in a critical solenoid valve in the yellow control pod and that the blue control pod AMF [dead-man switch] batteries had insufficient charge."

Transocean disputes BP's analysis of the blowout preventer (for one thing, the blue pod wasn't retrieved until long after the accident). The results of the official forensic examination of the BOP have not been announced as this book goes to press. It seems likely that the truth about Macondo will always be clouded by uncertainties, errors, misinformation, and conflicts of interest among the principals. Much of this is being litigated, ferociously, and that makes it harder to find a consensus on what really happened.

One theory that has emerged over time, and which is a bit shocking given all that has been said and written to date, is that the blowout preventer wasn't really so dysfunctional after all. Maybe—says this theory—the BOP did essentially what it was designed to do and simply was overmatched by Macondo. Transocean believes that the dead-man function worked as designed, that the blind shear ram closed on the drill pipe. But by that point, the well was already roaring full blast. It's possible that the gasket-like rubber packers around the pinchers just couldn't seal the well—they were eroded away as the gritty fluid from the deep raced through the BOP. This might be the answer: The rubber couldn't handle Macondo's furious flow.

Perrow, again: "[G]reat events have small beginnings . . . Small failures abound in big systems; accidents are not often caused by massive pipe breaks, wings coming off, or motors running amok. Patient accident reconstruction reveals the banality and triviality behind most catastrophes."

A key element of the Perrow model is incomprehensibility: The actors in the drama cannot see, or do not see, the linkages that are dooming the system to failure.

Perrow writes: "In complex industrial, space, and military systems, the normal accident generally (not always) means that the interactions are not only unexpected, but are *incomprehensible* for some critical period of time."

This incomprehensibility goes with the territory in any technology, but it's exacerbated when the crucial hardware is in a remote place—low earth orbit, say, or the deep sea. When the oil industry drills in shallow water, the blowout preventer is often right there on the deck of the rig or platform. When the industry goes into deep water, the blowout preventer migrates to the seafloor, where it is out of reach. It may be visible in the camera of an ROV, but you can't walk up and poke it, can't fiddle with a three-inch kill line to see if it's clogged. Depth matters.

Viewed broadly, the Deepwater Horizon accident should have been expected. It did not come out of the blue, but out of the system of deepwater drilling. It was a Black Swan event insofar as it was anomalous, unexpected, throwing everyone for a loop—a crisis no one had been thoughtful enough to plan for in advance—but it was also the natural consequence of specific decisions and practices that made it possible and, over time, not so unlikely. The Macondo blowout was a bit like the financial crisis of 2008 that sent the US economy into a recession: Something that emerged from a complex system, one in which the participants could not possibly track every moving part or fully comprehend what they were witnessing. In the same way that the bundled financial "instruments" concocted by Wall Street became inscrutably complex, the technology of deepwater oil drilling had become so elaborate that no single individual could possess full clarity on the operation.

The presidential oil spill commission discovered that the Horizon disaster joined a long list of mishaps, some of them not well known, in the offshore industry of the Gulf of Mexico. The presidential commission's staff compiled a list of thirty-three major offshore incidents in the gulf, going back to the 1979 Ixtoc I blowout in the Bay of Campeche. Four involved BP, and two involved Arco, which BP had purchased in 2000.

That was the general picture of the Deepwater Horizon disaster. But there were still some specifics to be figured out.

A Dangerous Business

On Sunday, October 3, I flew into New Orleans, the plane riding a cold front that blasted away a summer's worth of humidity and put a sparkle in the landscape. There were more Coast Guard hearings the next day, but I had some spare time and made a run down to Grand Isle.

At Port Fourchon, the oil industry boomtown, a roadblock remained at the entrance to the beach. I decided to try my luck farther east at Elmer's Island, but that was closed too, with a trooper standing guard. Finally, at Grand Isle, I found a place to walk on the beach. A few families braved the chill air. One man swam in the gulf. A boy with one leg played in the sand. He rolled like a tumbleweed, laughing.

At the jetty on the west end of the island, another large family swam in the shallows. There was no sign of oil, so far as I could see, though the beach is so dark—almost black in the wet spots—that it might be hard to spot at first. I got on my hands and knees and studied the beach sand. It's weirdly black. Was that from oil? No, just Mississippi River sediment, the schmutz of the continent.

I think.

On my knees staring at sand: the intrepid journalist getting to the bottom of things.

I went to the east end of the island, where the beach was still closed, surrounded by red plastic fencing. I came across an uprooted palm tree that lay in the sand with police crime tape swirling around the fronds, as if the tree had been murdered.

At Artie's Sports Bar, three women played the electronic poker machines while a few patrons drank beer at the bar. There was no oil spill paraphernalia anywhere except for a funny picture of Barack Obama in a Barney Fife sheriff's deputy cap, grinning like *The Andy Griffith Show*'s Don Knotts, with a caption, "That oil spill sure is somethin', Andy!"

The guy next to me at the bar worked on an oil platform just off Grand Isle. The moratorium didn't affect him. I thought of pulling out my notebook and grilling him, but here's a secret of journalism: It's more fun when the story is hot and the presses are hungry for copy and the ordinary thoughts of ordinary people suddenly are jacked up to extraordinary significance. Back in May, I had been ducking into these places, grabbing quotes so fast that it was almost like a theft, and filing them immediately to add to Web stories being updated repeatedly throughout the day. This had been the front line of the war. Now it was almost all the way back to being the Island That Time Forgot. My notebook stayed in my pocket. We talked football. On the TV behind the bar, the Redskins squeaked out an improbable victory over the Eagles.

The spill had hit us fast, and it left the same way.

On October 4 the Marine Board opened another session in the New Orleans suburb of Metairie, in the densely curtained second-floor Magnolia Room of a Holiday Inn. No more Radisson by the I-10 freeway—now we were at the Holiday Inn by the I-10 freeway. The government investigation had migrated about three exits. Rumor had it that the feds cut a deal to save $10,000; gone, suddenly, were the free coffee service and afternoon cookies. We were starting to rough it, precisely in keeping with the story vanishing from the front page.

There were fewer reporters than at the earlier hearings. But there were just as many lawyers and maybe a few more. The cleanup was winding down on the beaches, but the lawyers were just warming up.

On Thursday, October 7, we finally heard from Gregory Walz, the engineering team leader for the Macondo well. He had come into that position late, just a few weeks before the blowout. He seemed like classic middle management: portly, round faced, bald on top. This was not the guy in the executive suite, in case anyone was confused on that point:

> Q: Where did you sit in the office in relation to Mr. Hafle?
>
> A: He would be several cube sets over . . . there's all dividers. We're in cubicles.

Jason Mathews, questioning Walz, exhibited an incomprehensible chart showing the different levels and types of decisions involved in drilling a well and the different people responsible for each decision. To an outsider, the chain of command was confusing. Within BP, for example, the well team and the engineering team operate in separate, but parallel, lines of authority. Mathews tried to sort through this complex operation and figure out how BP bureaucratized engineering decisions and risk in general.

> Q: Can you please give me an example of what a tier-three decision would be in your opinion?
>
> A: Elimination of casing string, shoe release, and the side tracks versus fishing, the mud weight changes outside of the program guidelines, and then it gets into around changes and coordinate plans and subsurface issues.

Walz testified for many hours, and at no point did he stipulate any fundamental error of judgment, hesitating even to say that a cement bond log ought to be run in every such circumstance. He mentioned something I'd never heard: While the engineers tried to decide how to finish the well, and what design to use, they considered the option of permanently abandoning the troubled Macondo well. It didn't have to be a "keeper." They didn't have to convert it to a production well.

Walz insisted that safety was never compromised even though, according to information that investigator Mathews entered into the record, an employee's compensation at BP in some instances was influenced by the employee's ability to cut costs. According to Mathews, one BP employee documented on a spreadsheet that he'd saved the company $450,000 in a single year. Mathews revealed that the performance evaluations of the BP company men (well-site leaders on the rig) include a category, worth 15 percent of the overall assessment, called "Every Dollar Counts and Simplification." Of thirteen performance evaluations reviewed by the investigators, twelve employees had documented large amounts of money they had saved BP, typically upward of $100,000, Mathews said.

> Q: Do you feel that a prudent operator would require their employees in a performance measurement to reduce costs on a rig and basically, according to their own documents, meet weekly to figure ways to reduce costs?
>
> A: We are a business.
>
> Q: I understand that.
>
> A: And it's from that relationship as far as from a business, but you don't cross the line with safety and environmental.

Later in the morning, Brad Brian, one of the many attorneys on hand representing Transocean, grilled Walz about some of the emails that made references to the cost of certain actions.

> Q: Do you see where it says, where it talks about in the middle of the page, "The liner if required is also an acceptable option, but will add an additional seven to $10 million" in costs? Do you see that?
>
> A: Yes, sir.
>
> Q: And that was a factor in the decision-making process, wasn't it?
>
> A: I don't understand your question. I'm sorry.
>
> Q: Well, you put that—Somebody made a decision to put that in there because you thought it was a factor to be considered by the ultimate decision maker, didn't you?
>
> A: That was a contingency plan that we had in place, because at the time this was a—It was a contingency plan.
>
> Q: Did you think it was relevant or not, "yes" or "no," relevant or not to the decision to use the long string, that the liner would have cost up to ten million dollars more? Was that a relevant factor or not?
>
> A: The cost, no.
>
> Q: I'm sorry?
>
> A: The cost, no. It was not a factor on this, this decision.

Walz wouldn't budge on the cost-versus-safety issue. His testimony echoed something that Tony Hayward, who would soon step down as CEO, had said on September 15: "We have found no evidence in our assessment and investigation of this accident to suggest that costs were any part of how this occurred."

* * *

On November 17, 2010, the National Academy of Engineering released an interim report on the oil spill that made an observation that related directly to what Walz and others had said during testimony: "The often-made assertion at the MBI hearing that safety was never compromised suggests that the risks that are an inherent part of engineering processes in this and many other industries were not fully recognized."

BP and its contractors underestimated the risks of their operation. It's an obvious conclusion, but from National Academy of Engineering, it carried a great deal of weight. The academy also noted that the Macondo well team leader—John Guide, based in Houston—"was responsible for cost and schedule in addition to decisions affecting the integrity and safety of the well."

For the academy, it was a mistake to put a well team leader in a position of juggling efficiency and safety. Inevitably, the two goals will collide with each other. Doing that extra cement test? Takes time. Waiting around for parts to be delivered? Takes time. Should you speed things up by doing simultaneous operations (say, off-loading mud to a supply boat while doing a pressure test on the well) even if that complicates monitoring the well?

Of course decisions in the drilling business are made with cost in mind. If the only consideration was safety, they wouldn't drill oil wells in the first place.

It's dangerous.

BP did take steps to shake up its operation. In late September, preparing to take over the company, Bob Dudley announced that, among other internal changes, BP would create a new safety and risk unit, headed by Mark Bly, with authority to intervene in all technical activities. The company said in a press release that it would review how it "incentivises business performance, including reward strategy, with the aim of encouraging excellence in safety and risk management."

A couple of big names would vanish from BP in the weeks and months ahead. Andy Inglis, Hayward's protégé, gave one speech in late September espousing the lessons learned in the disaster response, and learned soon thereafter that he was out at the company. The spill happened on his watch and his ouster surprised no one. As one colleague put it: That's how the world works. And Doug Suttles, one of the company's most visible people during those jangling early weeks of the disaster, didn't stick around, either. In January 2011, Dudley announced that Suttles would be retiring after twenty-two years with the company.

The Obama administration's oil drilling conundrum came to a resolution, more or less, on December 1, 2010, when Interior Secretary Ken Salazar announced a partial retreat from the president's position of late March, pre-blowout. The administration unveiled a five-year drilling plan that extended the ban on drilling in the eastern Gulf of Mexico and along the Pacific and Atlantic coasts. The main concession to the oil industry was the opening of shallow water areas off the coast of Alaska, where Shell had been poised to start drilling. Some environmentalists were disappointed with the Arctic portion of the decision—the polar bears would still be imperiled. But the petroleum industry was dismayed to learn that, once again, it would be bottled up for years in the western Gulf of Mexico. Moreover, the administration had come up with a stringent set of new regulations for deepwater drilling. The moratorium had been lifted, but there wasn't any new exploratory drilling going on in the deep water.

Climate change legislation, meanwhile, was going nowhere. The original Democratic legislation, passed easily in the House, had stalled in the Senate, and then the Republican landslide in the November 2010 congressional elections ensured that little would happen on the carbon front in the near future.

On January 11, 2011, the presidential commission delivered its report, *Deep Water: The Gulf Oil Disaster and the Future of Offshore Drilling: Report to the President,* detailing nine separate risk-based decisions by BP and its contractors that might have increased the danger of a blowout. The commission warned that, without changes in the industry—for example, a new generation of blowout preventers, capping stacks, top hats, ROVs, pressure gauges, seismic and acoustic detectors, well-monitoring equipment, and so on—this kind of blowout will happen again. The commission recommended reforms in government regulation, as well as the creation of an industry-based safety institute.

The technological response to the blowout may never be ranked in the public mind with Apollo 13 as an improvisational engineering triumph, but the commission, having studied closely how the well was capped and killed, saw much to praise in the efforts of the engineers and scientists:

> The operation of numerous ships and remotely operated vehicles, in close proximity to one another and to gushing hydrocarbons, with no significant accidents was a credit both to BP's controls and to the Coast Guard and MMS officials who reviewed BP's procedures. BP's efforts to develop multiple source control options simultaneously were herculean. And the speed with which government scientists, with little background in deep-sea petroleum engineering, established meaningful oversight was truly impressive. The hundreds of individuals who spent the spring and summer of 2010 working to stop the spill, under enormous pressure and conditions of great uncertainty, have much in which to take pride. These remarkable efforts were necessary, however, because of a lack of advance preparation by industry and government.

That matched what I saw, too, all the way down to the final caveat. *Herculean* is a good word, one you don't hear much these days. But none of this should have been necessary. This shouldn't have happened. When it did, they weren't ready for it.

And so we applaud the way they climbed out of the hole, but we must pointedly note that they were the ones who dug it to begin with.

Epilogue

An Engineered Planet

What we thought we were seeing in the summer of the spill was a worst-case scenario, a blowout as bad as a blowout can be, leading to "the worst environmental disaster America has ever faced," as President Obama put it in his June 15, 2010, Oval Office address to the nation. But it's hard to argue that it was worse than the Dust Bowl, or worse than the slow-motion environmental disasters that are chronic rather than acute—the draining and poisoning of the Everglades, the suburban sprawl that has consumed once pastoral landscapes, to name just a couple of things that come quickly to mind.

It may be years before we know the full environmental impact of the BP oil spill, an assessment complicated by the depth and complexity of the ecosystems involved. The government's *Oil Budget* showed tens of millions of gallons of oil unaccounted for. It's still out there, somewhere, of unknown portent. Some of the compounds in the oil, the asphaltenes, could form tar mats that persist for many years, as has happened on reefs near the Ixtoc I blowout of 1979.

The only thing that is likely to last longer than the asphaltenes is the litigation.

The Macondo blowout may turn out to have been something of a warning shot for offshore drilling. The truly worst-case scenario is a multi-billion-barrel reservoir of hydrocarbons bleeding out into a fragile body of water; a major blowout in the Arctic could be particularly catastrophic. What we learned in 2010 was that such a thing isn't far-fetched. Steve Chu and his fellow scientists were right to fret about the worst-case scenario. The next time something like this happens, the country will want someone like Chu on task, the designated worrier.

Even if there's not another deepwater oil well blowout anytime soon, there will be *something* that happens, something awful and unexpected, that involves the failure of a complex technology. It could happen in outer space, at the bottom of the sea, in a nuclear power plant, on the electrical grid, or somewhere in the computer infrastructure that networks the planet. The Deepwater Horizon tragedy is a reminder of how little most of us know about modern technology. We don't know how anything works. We don't really think about the source of the electrons that somehow illuminate the lightbulb. We don't know where the gasoline comes from, exactly. No one can fix his or her own car anymore. We don't know where our food comes from, or even, in the case of certain processed substances, what our food is made of. (Petroleum, probably.) So it's only natural that, when there's a complex technological disaster, we can't render any kind of intelligent judgment about what ought to be done or who to hold accountable. We struggle to siphon the good information from the Internet's vast reservoir of rumor and nonsense and balderdash.

We're all sort of lost in our own society. We have made a world that is hard to understand. The modern condition is to spend a lot of time wondering what in the hell we're looking at. What is this thing? What's it doing? How does it work? What will happen next? What should I do?

In his masterpiece *One Hundred Years of Solitude,* Gabriel García Márquez describes Macondo as a place not yet emerged from the age of magic:

> At that time Macondo was a village of twenty adobe houses, built on the bank of a river of clear water that ran along a bed of polished stones, which were white and enormous, like prehistoric eggs. The world was so recent that many things lacked names, and in order to indicate them it was necessary to point.

That's how we still live, many of us. We don't know the names of things. We are surrounded by doohickeys and thingamajigs.

The irony is that we're inhabitants of a planet that is becoming increasingly engineered. The engineers are brilliant and creative, and most of us have little appreciation for what they do, so deftly is their handiwork woven into our daily lives. Nor do we adequately appreciate the labor of those who keep this highly engineered world running. They get no glory. Some of them spend much of their lives offshore, out of sight, unknown to the rest of society until one day maybe a tragedy puts their name in the paper. As in: Jason Anderson, Aaron Dale Burkeen, Donald Clark, Stephen Curtis, Gordon Jones, Roy Wyatt Kemp, Karl Kleppinger Jr., Blair Manuel, Dewey Revette, Shane Roshto, and Adam Weise.

We need to remember that sometimes bad things happen to complex systems, that gremlins roam the earth. Things go wrong: Count on it. The engineered planet challenges all of us to be a little smarter, to pay more attention. We need to learn the jargon, understand the risks.

There will be more black plumes. There will be other fires on the horizon. Low-probability, high-consequence events are made all the more devastating, potentially, by the scale and sophistication of modern technology. The human race is gambling that an engineered planet can be made sustainable, nuclear weapons controlled and managed, crops and livestock genetically modified, machines deftly crafted on the nanometer scale, the electrical grid revamped to be more highly networked and "smart," and perhaps the entire planet itself "geoengineered" to combat climate change. As we go down this technological path, we will count on complex systems to work correctly. We will assume that someone smart is in charge, looking over our world, protecting us. We will imagine a world full of blowout preventers that will actually prevent blowouts.

Here's the thing: Usually the technological magic works. Usually nothing terrible happens.

Usually.

As we grope our way forward, we can develop a few rules. Such as:

When doing something risky, remember that risk builds like plaque.

Make sure that your backup plan is really in back and won't get blown up out front along with your plan A.

Remember that low-probability, high-consequence events become more likely given enough time and opportunity. This is why you tell your teenager that you don't want her running around the party district at two in the morning. Sure, she'll probably stay out of trouble tonight, but it's her entire *adolescence* that you have to worry about.

Measure your misery. Don't shy away from knowing precisely how badly you're screwed. An appropriate response to a problem requires knowing the scale of the calamity. There's a reason you say to the doctor, "Tell me straight how bad this is."

Keep the fixers away from the talkers. Don't expose the engineers to any political shenanigans, media madness, or public outrage. Once a certain level of urgency is attained, the problem won't get solved more swiftly by knowing that everyone hates you.

Don't overpromise. Follow the Thad Allen rule: underpromise and overdeliver.

And, finally, the most important lesson: Keep your wits about you. It is extraordinarily unlikely that the disaster you are dealing with is qualitatively worse than the many calamities that human beings have survived to this point. In fact, it's probably not as difficult as any number of challenges that people have overcome, from wars to famines to pestilences to floods to storms to earthquakes. People survive, rebuild, thrive. The strange thing about Armageddon is that it never actually happens. So don't panic. The problem will be solved. Might not be pretty, but it'll get done.

One day in Houston, a petroleum engineer told me that the industry didn't make enough adjustments as it ventured into deep water. The oil industry thought it could keep doing what it had been doing. The deep, however, was a completely different world.

I said it was like the frog—

He nodded.

—"It's the frog in the boiling water."

You know that story. There's a frog sitting in a pan of cold water that is gradually brought to a boil. The frog never realizes that the time has come to jump to safety. Result: boiled frog.

But the oil industry will continue to drill in deep water. That's where the oil is. Among the companies that are particularly gung-ho about deepwater drilling: BP.

If anything, BP has become even more determined to drill in the most challenging environments. On January 15, 2011, BP chief executive officer Bob Dudley announced that the company had reached an agreement with the huge Russian oil company Rosneft to drill off the coast of Siberia in the Arctic Ocean. The deal includes a stock trade and gives the Russians access to BP's expertise. In exchange, BP can drill in huge tracts of the oil-rich Russian Arctic with no interference from US regulators.

At the end of January, BP announced that it had lost $4.9 billion in 2010 and estimated the cost of the gulf oil spill at $40.9 billion. But in a statement to shareholders, the company announced that it was resuming the dividend it had suspended in June. The company stressed that it had taken internal steps to increase the safety of operations. Speaking to reporters, Dudley said the new corporate strategy would be built around exploratory drilling, including in deep water. According to a report in the English newspaper the *Guardian,* a reporter asked Dudley if the emphasis on Arctic and deepwater drilling was appropriate given what had happened in the Gulf of Mexico. Dudley answered that the company should not retreat from deepwater technology, but rather should apply the lessons learned from the Macondo disaster.

To do otherwise, Dudley said, would be "irresponsible."

Notes on Sources

Endnotes and citations, along with links to sources, a bibliography and further information about the Deepwater Horizon disaster as new material becomes public, can be found online at www.aholeatthebottomofthe sea.com.

The reporting for this book relied on both public and unpublished documents. I obtained 19,218 pages of unpublished, previously undisclosed government emails that provided a real-time look at the management of the crisis and the often tense relationship between the government and BP. While covering the Marine Board of Investigation hearings I obtained (as did other members of the news media) hundreds of pages of still-unpublished documents. These include statements from survivors of the blowout, notes by BP investigators probing the accident as part of the company's internal investigation, investigative documents prepared by the Marine Board, Macondo well diagrams, Halliburton cementing reports, and internal BP emails. I interviewed many principal figures in the disaster response, including Thad Allen, Steve Chu, Tom Hunter, Marcia McNutt, Bob Dudley, Richard Lynch, and Kent Wells.

The scene in Chapter One in which I describe the view of Louisiana from the helicopter is based on my identical trip to the disaster site, on an identical helicopter, two months later. The description of the events on the rig on April 20 is based on testimony given before the marine board, as well as internal BP documents. I attended the July, August, and October sessions of the Marine Board. The Marine Board has an excellent and searchable website, including video and transcripts, at www .deepwaterinvestigation.com.

The presidential Oil Spill Commission produced an excellent series of "working papers" and then a final report in January 2011, plus an invaluable "chief counsel's" report that provided fresh material that I have used in Chapter 12. Several events in my book were first unearthed by the commission. I learned of USGS scientist Paul Hsieh's role in the response through a commission working paper titled "Stopping the Spill: The Five-Month Effort to Kill the Macondo Well." My account of President Obama's actions regarding the "sand berms" proposed by Louisiana officials is also based on a commission working paper, "The Story of the Louisiana Berms Project." Both papers and the commission's final report can be found at the commission's web site: www.oilspillcommission.gov/. The National Academy of Engineering issued an interim report on the disaster, which can be found at www.nae.edu. BP's "Bly report," though not a disinterested document, provides abundant technical data and well diagrams, in addition to the company's interpretation of the blowout. It can be found at www.bp.com.

In addition to the work of my colleagues at *The Washington Post* (see the Acknowledgments), journalists with many other news organizations provided outstanding coverage of the disaster and aftermath; I benefited in particular from the work of the Associated Press, *The Wall Street Journal, The New York Times, The Houston Chronicle,* and *The New Orleans Times-Picayune.* During the oil spill I regularly checked the technical discussions at the online forum The Oil Drum (www.theoildrum.com/).

This remains a dynamic event, one that is still under intensive investigation and subject to much litigation, and I expect the story will evolve in the months and years ahead. I'll provide updates online for readers wanting to know more. Readers should contact me at joel@joelachenbach.com. Please let me know if I can post your comment on the book's website. Readers may also want to visit my *Washington Post* blog, Achenblog, at www.washingtonpost.com/achenblog.

Acknowledgments

On Friday, April 30, 2010, *Washington Post* national editor Kevin Merida asked me to come into the office over the weekend and help out with the coverage of the Deepwater Horizon disaster. So began an immersion into a story that gave me the chance to be part of a wonderful team of writers and editors.

Steve Mufson, the *Post*'s tireless energy reporter, led the way day in and day out. David Fahrenthold is a gifted storyteller who plunged into the thick—the goo, literally—of the crisis on the beaches of the gulf. Environmental reporter Juliet Eilperin worked her countless sources to keep us in the loop on every breaking development. The four of us learned to mimic each other in print and finish each other's sentences when devising the daily battle plan. Frances Sellers, my editor, somehow kept track of an ever-growing army of reporters and, along with Kathryn Tolbert, edited copy all day long and into the night, only to start it all up again before dawn the next day. Ann Gerhart played a crucial editing role in the early days of the crisis. Thanks also to Jill Bartscht, Marcus Brauchli, Charity Brown, David Brown, David Hilzenrath, Marisa Katz, Marc Kaufman, Madonna Lebling, Jeff Leen, Kevin Merida, Liz Spayd, Stephen Stromberg, Marilyn Thompson, and Katharine Weymouth.

My editor at Simon & Schuster, Alice Mayhew, suggested this book and guided it into print with her usual mastery. Roger Labrie once again provided indispensable editing, wisdom, and encouragement while putting up with the usual chaos and madness of an Achenbook.

I am blessed with loyal friends who happen to be among the very finest journalists in the country. Gene Weingarten and David Von Drehle rescued me time and time again, bucking me up, pushing me toward the finish line, and reminding me that a good story is not necessarily measured by the amount of copy devoted to blowout preventers and 3-ram capping stacks. Pat Myers helped me write in English: She's the world's greatest copyeditor in addition to being the increasingly famous Empress of the Style Invitational (if you don't know it, Google it—worth the trip). Michael Lewis jumped in with emergency storytelling advice.

I owe a great debt to researcher Lucy Shackelford, who helped me throughout this long process, tracking down facts and sources, catching mistakes, and being there for me in crunch time.

Jonathan Deason, lead professor of George Washington University's Environmental and Energy Management program, kindly invited me to be a visiting scholar in residence in the Department of Engineering Management and Systems Engineering. Thanks also to the department's chair, Julie Ryan, and to professor Michael Duffey for offering advice on the manuscript.

Thanks also to my longtime agent, Michael Congdon; to the rest of the S&S team, including Jonathan Karp, Rachel Bergmann, Tracey Guest, Philip Bashe, Rachelle Andujar, and Nancy Inglis; and to the many professionals who put up with my countless demands for information, including Dave Cohen, Karla Hubbard, Byron King, Greg McCormack, Tyler Priest, Charles Royce and Richard Sears.

I had many freelance editors, and owe special thanks to Michael Baker, Kimberly Baker, Lauri Menditto and Linda ReVeal, who offered priceless editorial counsel, encouragement and friendship. Frances Sellers, undeterred by flu, provided a surgical edit. Thanks also to Kevin Achenbach, Geraldine Brooks, Geoff Dawson, Meg Dawson, Marc Fisher, Kyle Gibson, Jody Goodman, Tony Horwitz, Walter Isaacson, Annette Larkin, Jennifer Leete, John Menditto, Emily Notestein (also known as Mom), Jim Notestein, Mark Patterson, Molly Phee, John ReVeal, Tom Shroder, Mit Spears, Angus Yates, Sissy Yates, and Tom Mann and his League of Extraordinary Gentlemen.

Mary Stapp took time out from teaching journalism and running our household to mark up the manuscript. To Mary and the Achenstapp girls, Paris, Isabella and Shane: Thank you for your love, support and patience.

A final shout-out to Ken Deffeyes, my college geology professor, who helped me understand the Macondo blowout and offshore drilling in general. He is a brilliant scientist with a literary gift and a great sense of humor. Thanks, Professor.

Index